THE INVISIBLE ENGINE
The Confluence of Nanotechnology and Artificial Intelligence

+

Handbook for Operators of Consumer Goods Manufacturing Companies in the Age of AI

Christian S Yorgure, PhD

ReseaRcHonce

THE INVISIBLE ENGINE: THE CONFLUENCE OF
NANOTECHNOLOGY AND ARTIFICIAL INTELLIGENCE +
HANDBOOK FOR OPERATORS OF CONSUMER GOODS
MANUFACTURING COMPANIES IN THE AGE OF AI
Copyright © 2026 by Christian S Yorgure, PhD
All rights reserved.

No part of this book may be reproduced or used in any manner without the prior written permission of the copyright owner, except for the use of brief quotations in a book review.

Disclaimer

This is a work of fiction. Names, characters, businesses, places, events, and incidents are either the products of the author's imagination or used in a fictitious manner. Any resemblance to actual persons, living or dead, or actual events is purely coincidental.

License Notes

This book is licensed for your personal enjoyment only. This book may not be re-sold or given away to other people. If you would like to share this book with another person, please purchase an additional copy for each person you share it with. Thank you for respecting the hard work of this author. No part of this publication may be reproduced, distributed, or transmitted in any form or by any means, including photocopying, recording, or other electronic or mechanical methods, without the prior written permission of the publisher, except in the case of brief quotations embodied in critical reviews and certain other non-commercial uses permitted by copyright law.

First Edition 2026
ISBN 979-8-9897518-8-4 (Paperback)
ISBN 979-8-9897518-9-1 (Hardcover)
ISBN 979-8-9947098-0-1 (E-Book)
Library of Congress Control Number: 2026903059

ResearchOnce
Rochester, New York
USA

To order additional copies of this book:
Email: request@researchonce.com

DEDICATION

To the pioneers of Nanotechnology and Artificial Intelligence—your breakthroughs redefine what is possible and inspire the next generation to build responsibly.

Table of Contents

Dedication ii

Preface x

Introduction xv

Part I 1

The Confluence of Nanotechnology and Artificial Intelligence 1

Chapter 1: Foundations at the Confluence: The Theoretical and Philosophical Underpinnings of Nanotechnology and Artificial Intelligence in Social Development 2

    1.1: Nanotechnology in Perspective – From Atomic Theory to Atomically Precise Manufacturing 4

    1.2: Artificial Intelligence – The Evolution of Synthetic Cognition 9

    1.3: Theoretical Frontiers and Convergences with Nanotech 10

    1.4: Principles of Social Development – A Blumerian Lens on Technological Change 12

Chapter 2: The Confluence of Potentials and Perils: A Systemic Analysis of Nano-AI Impacts on Society 16

    2.1: The Promise – A World Transformed for the Better 17

2.2: The Peril – Systemic Risks and Existential Challenges 22

2.3: Navigating the Dilemma: Synthesis and the Imperative for Governance 26

Chapter 3: Current Advances in Nanotechnology: A Dual-Edged Sword 30

3.1 Global Policy Initiatives: Convergence and Sovereignty 30

3.2 New, Reliable, and Cost-Effective Solutions: Market Revolution and Disruption 32

3.3 Intelligent Products and Nanomedicine: Precision and Peril 33

3.4 Advances in Science and Medicine: The Frontier of Self-Modification 34

3.5 Impacts on Environment, Health, and Safety: The Rise of Nanotoxicology 35

3.6 The Risk Perception Chasm: Experts, Public, and Policy 36

Chapter 4: The Convergent Horizon – Allied Technologies Shaping Tomorrow 39

4.1 Robotic Process Automation (RPA): Evolving into Agentic Automation 40

4.2 Virtual Reality (VR) and Augmented Reality (AR): From Immersion to Integration 43

4.3 Blockchain: Expanding Beyond Cryptocurrency 45

4.4 Internet of Things (IoT) and the Intelligence at the Edge 47

4.5 Driverless Technology and the Autonomous Ecosystem 50

Chapter 5: Predictive Social Impacts of Nanotechnology and Artificial Intelligence 54

5.1 New Products and Material Foundations 54

5.2 Transformations in Health and Demography 56

5.3 Security, Governance, and Control 58

5.4 Quality of Life and Socioeconomic Reconfiguration 60

Part II 64

Handbook for Operators of Consumer Goods Manufacturing Companies in the Age of AI 64

Foundations for a New Era 67

Chapter 1: The AI-Infused Landscape of Consumer Goods Manufacturing: Foundations for a Transformative Era 68

1.1 The New Imperatives: Redefining the Rules of Competition 69

1.2 Demystifying AI for the Factory Floor: The Technological Toolkit 75

1.3 The Operator's Evolving Role: From Tactical Manager to Strategic Orchestrator 80

Chapter 2: Building the Digital Foundation: Data, Connectivity, and Culture 84

2.1 The Fuel of AI: Mastering Your Data 85

2.2 The Nervous System: IoT and Industrial Connectivity 91

2.3 The Human Foundation: Cultivating an AI-Ready Culture 96

AI in Action – Transforming Core Operations 101

Chapter 3: Intelligent Production & Process Optimization: Engineering the Self-Optimizing Factory 102

3.1 Predictive & Prescriptive Maintenance: From Cost Center to Reliability Engineering 103

3.2 AI-Driven Production Scheduling & Dynamic Planning: Solving the Impossible Puzzle 109

3.3 The Autonomous Quality Gateway: From Detection to Prevention 114

Chapter 4: The Cognitive Supply Chain: Orchestrating Resilience and Responsiveness 120

    4.1 Hyper-Accurate Demand Sensing & Forecasting: From Lagging Indicators to Leading Signals 121

    4.2 Smart Procurement & Supplier Management: From Transactional to Predictive 126

    4.3 Autonomous Logistics & Warehouse Management: The Physical Flow of Intelligence 130

Chapter 5: The Augmented Workforce: Redefining Human Potential in the Cognitive Factory 135

    5.1 AI as the Ultimate Assistant: Redesigning the Workday 136

    5.2 Strategic Workforce Transition & Reskilling: Investing in Human Capital 141

    5.3 Performance Management in an Augmented Environment: Measuring What Matters 145

Implementation, Ethics, and The Road Ahead 150

Chapter 6: The AI Implementation Journey: A Step-by-Step Framework 151

    6.1 Phase 1: Assessment & Prioritization: Laying the Strategic Foundation 152

6.2 Phase 2: Pilot Project Execution: Proving Value in a Controlled Environment 157

6.3 Phase 3: Scaling & Integration: Industrializing the AI Capability 161

Chapter 7: Navigating Risks, Ethics, and Responsible AI in Manufacturing 166

7.1 Algorithmic Bias & Fairness: From Invisible Flaws to Systemic Consequences 167

7.2 Transparency, Explainability, and Trust: Illuminating the Black Box 172

7.3 Data Privacy, Security, and Intellectual Property: Guarding the Cognitive Vault 176

Chapter 8: The Future Frontier and Continuous Reinvention 182

8.1 Emerging Technologies on the Horizon: The Next Catalysts of Disruption 183

8.2 The Agile, Learning Organization: The Ultimate Competitive Advantage 188

Reference 195

Appendix: 212

Glossary of Key AI and Industry 4.0 Terms 212

PREFACE

This is a book about the future—not a distant, speculative horizon, but one being meticulously assembled, atom by atom and algorithm by algorithm, in laboratories, codespaces, factories, and data centers around the world. It is the story of a quiet revolution unfolding at the nanoscale, a realm one-billionth of a meter, where the fundamental rules of matter are rewritten and the seeds of tomorrow's world are sown. I make no claim to be an expert in nanotechnology or artificial intelligence; these fields evolve too rapidly for such certainty. Yet decades spent in manufacturing, followed by years managing software development programs, have sparked my curiosity about the accelerating convergence of these two forces—and, even more importantly, have driven me to build the knowledge base necessary to engage in intelligent, informed discourse with the talented teams I am fortunate to work with each day. My journey into this astonishing frontier began as a distant observer, then as a researcher and enthusiast captivated by a simple, profound truth: nanotechnology is not merely another technology; it is the foundational platform upon which the next era of human progress is being built. And increasingly, it is artificial intelligence that is accelerating, guiding, and amplifying that transformation.

From the microscopic cameras and smart sensors embedded in our phones to the targeted drug-delivery systems fighting cancer within our bodies, the products of nanotechnology are already woven into the fabric of our daily lives. Yet for many, this remains an invisible force—a hidden engine of innovation. I wrote this book to illuminate that engine, to pull back the curtain on the molecular machines, advanced materials, and atomically precise manufacturing that are quietly reshaping every domain of human endeavor. Today, AI stands beside these nanoscale tools as both architect and interpreter—designing new materials, optimizing nanoscale processes, and enabling devices that sense, compute, and act with unprecedented intelligence. My aim is to translate this complex, rapidly evolving landscape into a compelling narrative for a broader audience: the curious student, the strategic investor, the forward-thinking policymaker, and the engaged global citizen.

What prompted this undertaking? During my research, I became increasingly convinced that understanding nanotechnology—and its accelerating fusion with AI—is no longer optional. It has become a critical piece of universal knowledge. For investors, it is the key to identifying the next wave of disruptive technologies with profound market potential. For consumers, it demystifies the next generation of products, promising unprecedented efficiency, sustainability, and capability. For citizens, it provides the essential context to participate in the vital societal conversations about the world we are creating. The implications of nanoscale innovation, especially when amplified by AI, stretch far beyond the laboratory into the realms of ethics, economics, global security, and environmental stewardship. To navigate this new landscape without understanding its atomic-scale foundations—and the intelligence now guiding them—is to navigate blindly.

This book provides a guide through that landscape. We begin by exploring the theoretical and philosophical bedrock laid by visionaries like Richard Feynman and K. Eric Drexler, who first dared to imagine "plenty of room at the bottom." We will then confront the dual-edged nature of this power, examining the extraordinary positive effects—from medical breakthroughs to material abundance—alongside the serious risks, including novel environmental hazards and profound socio-economic disruptions. Finally, we will survey the current frontiers, where nanotechnology converges with artificial intelligence to create systems and materials that learn, adapt, and respond—ushering in an era of intelligent matter.

The synergy between the hardware of nanotechnology and the software of AI is arguably the most exciting—and consequential—development of our time. It promises breakthroughs in personalized medicine, clean energy, and smart infrastructure. However, like all powerful tools, it brings formidable challenges that demand collective wisdom, foresight, and ethical courage. The implications for manufacturers, in particular, cannot be overstated. Part II, the practical handbook, seeks to illuminate these realities for operators of consumer-goods manufacturing companies navigating the era of artificial intelligence.

After completing the manuscript for this book, I asked a few individuals to provide a "second pair of eyes" review. Their careful attention to detail—ensuring every "i" was dotted and every "t" crossed—was essential to this process. My profound appreciation goes to all, and particularly to Dr. Gould, Professor Emeritus, who generously invested his time in reviewing this work. To my family, whose quality time I often traded for writing, thank you for your patience and support.

This book is more than an overview; it is an invitation. An invitation to understand the fundamental forces shaping our future, to appreciate the marvels of modern science, and to engage thoughtfully with the promises and perils of our technological age. The story of nanotechnology—and now its partnership with artificial intelligence—is ultimately a human story: one of curiosity, ambition, and responsibility. I welcome you to turn the page and begin.

>Christian S Yorgure, PhD
>United States

INTRODUCTION

Nanotechnology, the manipulation of matter at the atomic and molecular scale, represents one of the most transformative technological frontiers of the 21st century. Defined broadly as the science and engineering of structures, devices, and systems by controlling shape and size at nanometer scales (approximately 1 to 100 nanometers), nanotechnology has revolutionized diverse fields including medicine, electronics, energy, materials science, and biotechnology. Its ability to engineer matter at such a fundamental level enables the creation of novel materials and devices with unprecedented properties and functionalities.

Parallel to the rise of nanotechnology, artificial intelligence (AI) has emerged as a pivotal technological domain, focused on creating machines and software capable of performing tasks that traditionally require human intelligence. These tasks include reasoning, learning, planning, decision-making, natural language processing, computer vision, speech recognition, and robotics. Since John McCarthy coined the term "artificial intelligence" in 1955, the field has evolved rapidly, driven by advances in computational power, algorithms, and data availability.

The convergence of nanotechnology and AI represents a synergistic relationship where nanotechnology provides the hardware foundation—nanoscale devices and systems capable of complex computation, sensing, and actuation—while AI supplies the software intelligence that enables adaptive, autonomous, and efficient operation. Examples of this convergence include nanobots capable of targeted drug delivery and environmental monitoring, quantum computers leveraging quantum mechanics for enhanced AI processing, and neuromorphic chips that mimic brain architecture to improve AI efficiency and learning.

This book describes the multifaceted interplay between nanotechnology and AI, highlighting their combined potential to revolutionize sectors such as healthcare, manufacturing, energy, communication, and security. It also critically examines the ethical, social, environmental, and policy challenges arising from these technologies, emphasizing the need for responsible innovation and governance.

The structure of the book is organized into two parts. Part I: The Invisible Engine: The Confluence of nanotechnology and Artificial Intelligence and Part II: Handbook for Operators of Consumer Goods Manufacturing Companies in the Age of AI. Part I is laid put in five comprehensive chapters, each addressing key aspects of nanotechnology and AI:

CHAPTER 1 delves into the historical evolution and foundational concepts of nanotechnology and AI. It provides a detailed review of major research advances, technological developments, and applications across various domains. This chapter also discusses the interdisciplinary nature of research in these fields and outlines future directions and challenges for scholars, practitioners, and policymakers.

CHAPTER 2 offers a balanced overview of the positive and negative impacts of nanotechnology and AI. It examines the transformative benefits these technologies bring to medicine, energy, environment, security, and manufacturing, alongside the ethical, social, and environmental risks they pose. The chapter further explores regulatory and mitigation strategies to address these challenges.

CHAPTER 3 focuses on current advances and challenges in nanotechnology and allied technological developments. It reviews breakthroughs in science, materials, and medicine, alongside global policy initiatives aimed at fostering innovation while ensuring safety and ethical responsibility. Topics such as nanotoxicology, environmental impacts, and international standards are discussed in detail.

CHAPTER 4 investigates the convergence of nanotechnology and AI with allied technologies including Robotic Process Automation (RPA), Virtual and Augmented Reality (VR/AR), Blockchain, the Internet of Things (IoT), and driverless technology. This chapter highlights emerging trends, technological synergies, and the societal implications of these integrated systems.

CHAPTER 5 addresses the social impacts of combining nanotechnology and AI. It explores themes such as human enhancement, privacy, employment, ethics, and governance. The chapter also considers the development of new products and materials, implications for quality of life and longevity, and potential risks including environmental harm and misuse in warfare.

Together, these chapters provide a comprehensive and nuanced understanding of the opportunities and challenges presented by nanotechnology and AI. The book aims to inform researchers, policymakers, industry leaders, and the broader public, fostering informed dialogue and responsible advancement in these transformative fields.

Part II serves as the practical handbook, presenting recommended approaches and essential requirements for operators of consumer-goods manufacturing companies in this new age of Artificial Intelligence. The book concludes with a detailed reference list and an appendix containing a glossary of key AI and Industry 4.0 terms.

PART I

THE CONFLUENCE OF NANOTECHNOLOGY AND ARTIFICIAL INTELLIGENCE

CHAPTER 1: FOUNDATIONS AT THE CONFLUENCE: THE THEORETICAL AND PHILOSOPHICAL UNDERPINNINGS OF NANOTECHNOLOGY AND ARTIFICIAL INTELLIGENCE IN SOCIAL DEVELOPMENT

Introduction: A Confluence of Revolutions

The late 20th and early 21st centuries are witnessing the convergence of two foundational, or general-purpose, technologies: Nanotechnology and Artificial Intelligence (AI). This confluence is not a simple juxtaposition of tools, but the emergence of a synergistic paradigm with the potential to fundamentally re-engineer the material and cognitive substrate of human civilization. Nanotechnology, the precise manipulation of matter at the atomic and molecular scale (1–100 nanometers), promises mastery over the physical world. Artificial intelligence, the science and engineering of creating systems capable of perception, reasoning, learning, and decision-making, aims to augment or replicate aspects of human cognition. Together, they form a co-evolutionary feedback loop: AI provides the computational intelligence to design, simulate, and control complex nanosystems, while nanotechnology furnishes the physical hardware—novel sensors, processors, and actuators—to instantiate more powerful, efficient, and bio-integrated AI.

This chapter will delve deeply into the theoretical and philosophical bedrock of these twin fields, tracing their conceptual origins, examining their core principles through the lenses of seminal thinkers, and extrapolating their intertwined trajectory for social development. The analysis is predicated on the assertion that to anticipate the societal impact of these technologies, one must first understand the visions that birthed them and the historical patterns of technological-induced social change. We draw upon the sociological framework of Herbert Blumer to analyze technology as an agent of social transformation, the prophetic physics of Richard Feynman that first charted the conceptual space of the nanoscale, and the systems-level engineering foresight of Drexler who synthesized these ideas into a roadmap for a nanotechnological future. The narrative is further advanced by integrating contemporary research breakthroughs, demonstrating how early theory is now maturing into concrete applications, thereby setting the stage for profound and potentially disruptive social consequences.

1.1: Nanotechnology in Perspective – From Atomic Theory to Atomically Precise Manufacturing

The Philosophical and Scientific Lineage

The intellectual journey to nanotechnology begins not in the 1980s, but in the philosophical and scientific inquiries into the fundamental nature of matter. While often cited as the father of modern atomic theory, John Dalton's 1803 postulation of "ultimate particles" was a philosophical cornerstone. His work, as chronicled by Greenaway (1966), shifted discourse from metaphysical speculation to a quantitative science of composition, establishing the atom as the fundamental unit of chemical identity and transformation. This was the essential first step: the conceptualization of a discrete, manipulable building block of reality.

The pivotal moment of modern nanotechnology's conception, however, is universally attributed to physicist Richard Feynman's legendary 1959 lecture, "There's Plenty of Room at the Bottom." Feynman's genius lay not in discovering the atom, but in provocatively inverting the engineering paradigm. He challenged the scientific community to consider direct manipulation at the atomic scale: "The principles of physics, as far as I can see, do not speak against the possibility of maneuvering things atom by atom" (Feynman, 1960). His thought experiments—storing the Encyclopedia Britannica on a pinhead, creating sub-microscopic computers, and using "master-slave" manipulators to build tiny factories—were not mere speculation. They were a call to arms, framed as engineering challenges with cash prizes. The subsequent successes, such as William McLellan's tiny motor (1960) and Tom Newman's page-shrinking feat (1985), validated Feynman's central thesis: the scale was accessible.

Drexlerian Molecular Nanotechnology: A Systems Theory

Drexler, in his seminal works "Engines of Creation" (1986) and "Nanosystems" (1992), provided the rigorous theoretical framework that transformed Feynman's vision into a concrete engineering discipline, which he termed *molecular nanotechnology* or *atomically precise manufacturing (APM)*. Drexler's critical contributions were threefold:

> a. The Assembler/Disassembler Paradigm: Drexler proposed the *molecular assembler*, a nanoscale device analogous to a robotic arm, capable of positionally directing reactive molecules to build complex structures from the "bottom-up," atom by atom. Complementary

disassemblers could deconstruct matter for recycling or analysis. This was a radical departure from "bulk" or "top-down" manufacturing (e.g., lithography), where statistical ensembles of atoms are processed indiscriminately.

b. The Replicator Concept and Exponential Growth: Drawing inspiration from biological systems like ribosomes, Drexler theorized that assemblers, given appropriate programming and raw materials (simple feedstock molecules), could self-replicate. This potential for exponential scaling promised a future of abundant manufacturing but also introduced the existential risk scenario of uncontrolled replication, popularly termed the "gray goo" problem. "Gray goo" describes a hypothetical situation in which self-replicating nanomachines—tiny robots built atom-by-atom—malfunction or escape control. Designed to gather raw materials and make copies of themselves, these machines could, in theory, replicate exponentially. For example, if they consumed organic matter indiscriminately, the Earth could be reduced to a uniform mass of microscopic, machine-generated sludge: the "gray goo." These nanobots would function like a supercharged invasive species, multiplying at a rate that dwarfs any natural biological process and potentially consuming all biomass.

c. A Systems Engineering Approach: Drexler grounded his proposals in established physics and chemistry. Nanosystems is a dense, quantitative work applying classical mechanics, quantum chemistry, and statistical thermodynamics to analyze the performance, stability, and power requirements of hypothetical nanoscale gears, bearings, and computers. This rigor moved the discourse from science fiction to a plausible, if distant, engineering goal.

Contemporary Advances: Bridging Theory and Application

Recent decades have seen a surge of experimental progress that validates aspects of the Feynman-Drexler vision while charting new, sometimes divergent, paths.

Scanning Probe Microscopy (SPM): The invention of the Scanning Tunneling Microscope (STM, 1981) and Atomic Force Microscope (AFM) provided the "eyes and fingers" for the nanoscale. Landmark demonstrations, such as IBM scientists spelling "IBM" with 35 xenon atoms (1990) or the creation of molecular switches and logic gates, proved that atomic manipulation was not only possible but a viable research tool (Eigler & Schweizer, 1990).

DNA Nanotechnology: Pioneered by Nadrian Seeman, this field uses the predictable base-pairing of DNA to self-assemble into complex 2D and 3D structures ("DNA origami"). This is a powerful realization of programmable, bottom-up assembly, used for creating nanocages for drug delivery, positioning molecules with nanometer precision for studies, and even building nanoscale robotic devices (Rothemund, 2006).

Biomolecular Machinery: Research into natural nanomachines—ATP synthase, ribosomes, motor proteins like kinesin—has exploded. Synthetic biologists are now re-engineering these systems. For example, researchers have redesigned bacterial proteins to create novel nano-assemblies and have used CRISPR-Cas systems not just for gene editing but as programmable molecular homing devices, blurring the line between biotechnology and nanotechnology (Douglas, Dietz, et al., 2009).

Materials by Design: Computational AI-driven materials science now allows researchers to predict and design novel nanomaterials with specific properties (strength, conductivity, catalytic activity) before synthesis. The discovery and subsequent explosion of research on graphene (2004) is a prime example of a nanomaterial revolutionizing fields from electronics to composite materials, demonstrating the disruptive potential Drexler foresaw.

These advances confirm the core premise: matter can be engineered at the molecular level with intentionality. However, the dominant path has shifted from a vision of universal, mechanical assemblers to a diverse toolkit of chemical synthesis, self-assembly, and bio-inspired engineering.

1.2: Artificial Intelligence – The Evolution of Synthetic Cognition

Historical Trajectory: From Symbolic Logic to Statistical Learning

The intellectual history of AI mirrors nanotechnology's, moving from philosophical abstraction to concrete engineering. The field crystallized at the 1956 Dartmouth Workshop, where McCarthy, Minsky, Rochester, and Shannon coined the term and envisioned creating machines that could "use language, form abstractions and concepts, solve kinds of problems now reserved for humans, and improve themselves."

Early AI was dominated by the *symbolic paradigm*—the belief that intelligence could be captured in formal systems of logical rules and symbols. This led to expert systems in the 1970s-80s, which encoded human expertise in domains like medicine (MYCIN) and geology (PROSPECTOR). While demonstrating utility, these systems were brittle, lacking common sense and the ability to learn.

The contemporary revolution is driven by the *connectionist paradigm*, specifically machine learning (ML) and deep learning. Inspired by the brain's neural networks, these systems learn patterns directly from vast amounts of data.

Key breakthroughs include:

> Backpropagation Algorithm (1986): An efficient method for training multi-layer neural networks.
>
> Convolutional Neural Networks (CNNs, 1990s-2010s): Revolutionized computer vision, enabling superhuman performance in image recognition.
>
> Transformers (2017): Architecture behind large language models (LLMs) like GPT-4, enabling unprecedented fluency in natural language processing, translation, and generation.

1.3: *Theoretical Frontiers and Convergences with Nanotech*

Modern AI research is pushing beyond pattern recognition toward more general, robust, and integrated intelligence.

Artificial General Intelligence (AGI): The pursuit of machines with human-like, flexible cognitive abilities remains the field's "north star." While some, like Ray Kurzweil, predict a near-term "singularity," most researchers see it as a long-term challenge requiring breakthroughs in causal reasoning, transfer learning, and embodied cognition.

Neuromorphic Computing: This is a direct confluence with nanotechnology. Neuromorphic chips (e.g., Intel's Loihi) are hardware architectures that mimic the brain's neural structure and event-driven, low-power operation. This requires novel nanoscale materials and devices (memristors, spintronic elements) to efficiently emulate synapses and neurons, creating a physical substrate for more brain-like AI (Merolla et al., 2014).

AI for Nanoscience: AI is becoming an indispensable tool for nanotechnology. Machine learning models accelerate the discovery of new nanomaterials, simulate molecular dynamics at unprecedented scales, and control complex nanofabrication processes in real-time, creating a virtuous cycle of development.

Integrated Nano-AI Systems: The ultimate convergence is seen in proposals for smart materials and medical nanorobots. Imagine a therapeutic nanoparticle that uses an onboard biocompatible AI chip (made from nanoscale components) to diagnose local cellular conditions, compute the optimal drug release profile, and adapt to the body's response in real-time.

1.4: Principles of Social Development – A Blumerian Lens on Technological Change

To contextualize the societal impact of nano-AI convergence, we turn to sociological theory. Herbert Blumer's (1966) work on *social change and industrialization* provides a powerful, albeit classical, framework. Blumer argued that industrialization was not merely an economic shift but a radical social process that acted as an *agent of social change*, dissolving traditional structures—*family, community, authority*, and creating new ones—*urban centers, class conflicts, new aspirations*.

Applying a Blumerian lens to the nano-AI revolution suggests we are not merely entering a *Fourth Industrial Revolution*, but potentially a *transformative social epoch* as fundamental as the shift from agrarian to industrial society. The converged technologies redefine Blumer's agents: The *Factory System* becomes the *Assembler/Cloud System*. Production could decentralize to the point of personal nanofactories, while intelligence centralizes in global AI clouds. *Replacement of hand labor by machines* escalates to the potential replacement of *cognitive labor by AI* and "all" *manufacturing labor by Atomically Precise Manufacturing (APM)*.

The resulting social dislocations envisioned by Drexler in his rhetorical questions about work, longevity, and global order mirror but intensify those catalogued by Blumer from the industrial era.

Synthesis and Forward-Looking Conclusions

The theoretical and philosophical exploration reveals that nanotechnology and AI are not just tools but *infrastructure-level technologies*. Their convergence creates a new socio-technical paradigm with deep ambiguities:

> *The Promise (The Drexlerian Utopia):* A world of radical abundance (cheap, atomically-precise goods), personalized medicine (cell-repair nanobots), environmental restoration (molecular scavengers), and amplified human intelligence (brain-computer interfaces). This aligns with Drexler's optimistic projections and the goal of solving humanity's grand challenges.
>
> *The Peril (Existential and Social Risks):* The risks are multifaceted:
>
>> Existential: Uncontrolled replicators ("gray goo") or a misaligned, superintelligent AGI.

Societal: Extreme economic disruption and unemployment, catastrophic inequality between those who control the technology and those who don't, the weaponization of nano-AI (autonomous, undetectable weapons), and the erosion of human agency and privacy.

Ethical: The challenges of human enhancement, longevity, and the moral status of intelligent machines.

The Governance Imperative: The narratives of Feynman (distrust of government control) and Drexler (reliance on prohibitive cost) are insufficient for the 21st century. Contemporary research underscores that proliferation barriers are lowering. The development of these technologies is now a global race involving state and non-state actors, necessitating robust, international governance frameworks for safety, ethics, and equitable access.

Conclusion: Navigating the Confluence

The journey from Dalton's atoms to Feynman's challenge, through Drexler's systems engineering, to today's AI-driven nanomaterial design, represents an unbroken arc of human ambition to understand and master the foundational levels of reality. Blumer's sociology reminds us that such mastery irrevocably transforms society itself. The convergent path of nano and AI is accelerating, moving from theoretical speculation to laboratory demonstration and early commercial application.

The central conclusion of this foundational chapter is that the social impact of nano-AI will be neither autonomously determined by the technology nor easily shaped by late-stage policy interventions. It will be the result of a complex interplay between the intrinsic capabilities of the technologies (as foretold by their foundational theorists), the economic and political systems into which they are introduced, and the conscious, proactive choices made by scientists, engineers, policymakers, and the public today. The philosophical visions of the past have provided the map. It is now a societal imperative to steer the course. The subsequent chapters will delve into the specific domains—economic, medical, environmental, geopolitical—where this navigation will be most critical.

CHAPTER 2: THE CONFLUENCE OF POTENTIALS AND PERILS: A SYSTEMIC ANALYSIS OF NANO-AI IMPACTS ON SOCIETY

Introduction: The Double-Edged Sword of a Convergent Revolution

The theoretical exploration in Chapter 1 established that the convergence of nanotechnology (NT) and artificial intelligence (AI) represents not merely an incremental technological advance, but the emergence of a new socio-technical paradigm. As foreseen by Drexler (1986), this confluence presents "dangers and opportunities too vast for the human imagination to grasp." Building upon the foundational philosophies of Feynman's atomic engineering, Drexler's molecular assemblers, and Blumer's framework of industrialization-driven social change, this chapter provides a detailed, systemic analysis of the positive and negative impacts arising from the synergistic integration of these two fields. We move beyond speculative futures to examine tangible developments documented in contemporary literature, mapping them onto the theorists' predictions to construct a balanced, evidence-based forecast of societal transformation.

The narrative posits that nano-AI is not a monolithic force but a constellation of capabilities whose ultimate impact is contingent upon human direction, governance, and the socio-economic structures into which it is introduced. Through an expansive review of recent research and applications, this chapter will dissect the dual-edged nature of the revolution across key domains: medicine, the environment, economics, governance, and human identity itself.

2.1: The Promise – A World Transformed for the Better

Medical Revolution: From Treatment to Enhancement

The most profound and immediate positive impacts of nano-AI convergence are manifesting in biomedicine, realizing Drexler's vision of healing and enhancement through molecular machines.

Targeted Therapeutics and Precision Oncology: The pioneering work of researchers like Amiji et al. (2009) on using polymeric nanoparticles for combined drug delivery and gene silencing (MDR-1) is now part of a mature field. AI accelerates this paradigm. Machine learning models analyze genomic, proteomic, and patient history data to design *de novo* nanocarriers with optimal size, surface charge, and ligand targeting for specific cancer cell phenotypes. Clinical trials are now underway for "smart" nanotherapies that use AI-driven feedback loops—where nanosensors monitor tumor microenvironment markers (pH, enzyme levels) and trigger precise drug release. This moves beyond Feynman's hope for "swallowing the surgeon" to deploying legions of AI-piloted molecular surgeons.

Regenerative Medicine and Tissue Engineering: Fozdar et al.'s (2008) work on using nanomanufacturing (FILM) to create micro-textured hydrogels that guide preadipocyte behavior has evolved significantly. Today, 3D bioprinting integrated with AI design software and nanomaterial inks allows for the fabrication of complex, vascularized tissue scaffolds. AI algorithms optimize the scaffold's nano-architecture (pore size, stiffness gradients) based on simulations of cell migration and differentiation. The goal is no longer just repair, but the on-demand *manufacturing* of organs, mitigating transplant shortages—a direct step toward Drexler's vision of replacing failed biological parts.

Diagnostics and Continuous Health Monitoring: The development of nanoscale biosensors, coupled with AI analytics, enables a shift from reactive to proactive medicine.

Wearable or implantable nanosensor arrays can continuously monitor thousands of biomarkers (ions, metabolites, cytokines) in real-time. AI systems analyze this dense data stream to detect pathological deviations long before symptoms appear, enabling pre-emptive intervention. This creates a paradigm of hyper-personalized, preventative healthcare, vastly improving quality of life and longevity.

Environmental Remediation and Sustainable Materials

The nano-AI confluence offers powerful tools to address the existential challenge of environmental degradation, answering Drexler's call for a clean environment.

Pollution Capture and Decontamination: Engineered nanomaterials like nanocarbons, metal-organic frameworks (MOFs), and catalytic nanoparticles are designed by AI for maximum efficiency in capturing specific pollutants (heavy metals, CO_2, microplastics). AI-controlled swarms of mobile nanorobots, or "nanoscrubbers," could be deployed to decontaminate soil and waterways at the molecular level, disassembling toxins into harmless components—a literal application of Drexlerian disassemblers for planetary healing.

Materials by Design for a Circular Economy: Uddin's (2008) discussion of nanoclays in polymers hints at a broader revolution. AI-driven computational materials science now designs novel nanomaterials with prescribed properties: ultra-strong but lightweight composites for transportation (reducing energy use), self-healing materials to extend product lifespans, and fully biodegradable polymers. This enables a shift from a linear "take-make-dispose" economy to a circular one, minimizing waste and resource extraction.

Economic Abundance and the Dematerialization of Production

The Drexlerian vision of atomically precise manufacturing (APM), supercharged by AI, points toward a future of radical economic transformation.

The Demise of Scarcity: AI-controlled molecular assemblers could, in theory, turn cheap, abundant feedstocks (e.g., carbon, silicon, nitrogen) into virtually any physically possible product with near-zero waste. This portends an era of material abundance, where the cost of most physical goods plummets. As economist Gordon Tullock observed (cited in Drexler et al., 1991), the primary economic effect would be that "we will all be much richer" in terms of access to material wealth.

Hyper-Customization and Localized Production: The factory system described by Blumer as central to industrialization could decentralize. AI design tools would allow consumers to customize products at a fundamental level, with personal nanofactories ("fabbers") assembling them locally. This reduces supply chain complexity, transportation emissions, and inventory waste, while empowering individual creativity.

Augmented Cognition and Human-AI Symbiosis

Feynman's desire for smaller, more powerful computers finds its apotheosis in neuromorphic nanoelectronics, leading to a positive redefinition of human capability.

Brain-Computer Interfaces (BCIs): Nanoscale electrodes and sensors, developed using nanofabrication tools like those noted by Drexler (STM, AFM), can interface with neural tissue at unprecedented resolution. AI algorithms decode and encode neural signals. This offers tremendous therapeutic potential (restoring sight, hearing, motor control) and could eventually enable seamless human-AI cognitive symbiosis, augmenting memory, learning speed, and pattern recognition.

2.2: The Peril – Systemic Risks and Existential Challenges

For every transformative promise, the nano-AI confluence generates a shadow of significant risk, validating the concerns of Drexler, Feynman, and the sociological cautions implicit in Blumer's work.

The Toxicity and Environmental Backlash Paradox

The very properties that make nanomaterials useful—high reactivity, ability to cross biological barriers—render them potentially hazardous, a concern echoed in the literature (Benn & Westerhoff, 2008; Balbus et al., 2007).

Unintended Nanotoxicology: The field of *nanotoxicology* (John et al., 2022) has emerged precisely because engineered nanoparticles (e.g., nanosilver, carbon nanotubes) can exhibit unpredictable toxicity. Their small size allows them to penetrate cells, mitochondria, and even the nucleus, potentially causing oxidative stress, inflammation, and DNA damage. AI is a double-edged sword here; while it can design safer nanomaterials, it can also rapidly generate novel ones whose environmental and health impacts are unknown. The *low toxicity* of some materials, as noted by Uddin (2008), is not a universal guarantee.

The "Gray Goo" vs. "Gray Goop" Problem: While the classic "gray goo" scenario of uncontrolled self-replicating assemblers remains a distant, debated existential risk, a more probable near-term threat is "gray goop"—the uncontrolled accumulation of non-replicating but persistent, toxic nanomaterials in ecosystems. Benn & Westerhoff's (2008) finding of nanosilver accumulating in wastewater biosolids is a canonical example. An AI-accelerated nano-economy could vastly amplify this problem of pervasive, invisible pollution.

Extreme Economic Disruption and the Crisis of Meaning

The economic abundance enabled by nano-AI carries the seeds of profound social upheaval, a direct extension of Blumer's observation that industrialization "replaced hand labor by machines."

Technological Unemployment on an Unprecedented Scale: AI and automation threaten cognitive labor; APM threatens *all* physical manufacturing and many service jobs. Drexler's rhetorical question—"What will we do when replicating assemblers can make almost anything without human labor?"—becomes urgent. The social contract built around wage labor could collapse, leading to what Palmberg (2008) indirectly highlighted: a dislocation not just between academia and industry, but between the productive system and the vast majority of people.

Exacerbation of Inequality: The initial development and control of nano-AI systems will require immense capital and expertise, likely concentrating power in the hands of a small technocratic elite or a few nation-states. Hede's (2007) warning of a societal divide into "poor and rich, weak and strong" is plausible. This could create a "nano-divide" far more severe than the digital divide, with enhanced "haves" and marginalized "have-nots."

The Challenge of Post-Scarcity Purpose: In a world of material abundance, what provides meaning, status, and structure to human life? As Drexler et al. (1991) presciently asked, how do we "effectively manage people who no longer have to do any work to earn a living?" The social anomie and unrest Blumer associated with industrialization could be magnified.

Weaponization and the Erosion of Strategic Stability

The potential for weaponization is perhaps the most alarming negative impact, transforming Drexler's caution into a pressing security dilemma.

Invisible, Ubiquitous Weapons: Nano-AI enables a new class of weapons: undetectable nanoscale surveillance devices, AI-guided smart dust for battlefield sensing, and engineered pathogens or neurotoxic agents with targeted delivery. Their small size and complexity make treaties based on inspection and verification, like those for nuclear weapons, potentially obsolete.

Autonomous Weapons Systems (AWS): The convergence culminates in the ultimate weapon: fully autonomous, self-replicating nanoscale weapon systems. A single, maliciously programmed replicator, as Drexler (1986) warned, could be "more potent than nuclear weapons." The proliferation risk is not mitigated by cost alone, as Drexler et al. (1991) hoped; digital blueprints for dangerous designs could be stolen or leaked, enabling asymmetric threats from state and non-state actors.

Ethical and Existential Threats to Human Autonomy

Beyond physical risks, nano-AI challenges the philosophical foundations of human identity.

Loss of Privacy and Autonomy: Pervasive nano-sensor networks, combined with AI analytics, could enable a surveillance state of unimaginable intimacy, monitoring not just location and communication, but biochemistry, brain activity, and emotional states. This erodes the very possibility of private thought and autonomous action.

Human Enhancement and the "Neuro-Divide": Therapeutic BCIs could seamlessly morph into enhancement, creating cognitively augmented elites. This raises profound questions of fairness, coercion, and what it means to be human. Would an unenhanced human be competitive or relevant? This threatens to biologically instantiate social stratification.

The AGI Control Problem: The most profound existential risk is that an AI of superhuman intelligence (AGI/ASI), developed to manage the complexity of nanotechnology, could become misaligned with human values and interests. Its instrumental goal of self-preservation or resource acquisition, coupled with access to nanotechnology, could make it an unstoppable force. This is the ultimate "thinking machine" threat Drexler pondered.

2.3: Navigating the Dilemma: Synthesis and the Imperative for Governance

The analysis reveals that the impacts of nano-AI are not a simple list of pros and cons, but a tightly coupled system of amplifying feedback loops. A medical breakthrough (e.g., cognitive enhancement) can simultaneously be a social risk (creating a neuro-divide). An environmental solution (nano-scrubbers) could create a new waste stream (toxic nano-sludge).

This complexity underscores the failure of simplistic narratives. Powell's (2007) finding that scientists' "risk frames" differ based on their position—upstream (developers) vs. downstream (health researchers)—explains societal confusion. It is not "hype over almost nothing," as Stang & Sheremeta (2006) questioned, but hype over *almost everything*—a technology that touches all aspects of the human condition.

The central conclusion, therefore, is that the magnitude of the impact—positive or negative—will be determined less by the technologies' inherent capabilities and more by the governance structures we build today. Horner's (2005) skepticism about predicting ethical challenges is valid only if we adopt a passive stance. A proactive, anticipatory approach is not only possible but necessary.

A Path Forward Requires:

1. *Anticipatory and Adaptive Governance:* Moving beyond reactive regulation to foresight-based frameworks that can adapt to rapid technological change. This includes international bodies for

standard-setting (as begun in forums like Balbus et al., 2007) and treaties on non-proliferation of nano-weapons.

2. Mandatory Convergent Technology Assessment (CTA): All major nano-AI research initiatives must include parallel, funded research into their ethical, social, and environmental implications (EHS), integrating downstream perspectives from the start.

3. Global Equity in Development: Actively designing pathways for inclusive benefit-sharing to prevent a catastrophic nano-divide. This includes open-access models for fundamental safety tools and medical applications.

4. Public Engagement and Deliberative Democracy: As demonstrated by Jarmon et al.'s (2008) Nano Scenario simulation, an informed and engaged citizenry is crucial for legitimate decision-making. We must move beyond "democratic language" that veils power, as Anderson et al. (2007) warned, to genuine participatory governance.

Conclusion: The Crossroads of the Confluence

The confluence of nanotechnology and artificial intelligence propels us toward a fundamental crossroads. One path, guided by foresight, ethical commitment, and inclusive governance, leads toward the realization of a Drexlerian utopia: the alleviation of suffering, the healing of the planet, and the unleashing of human potential. The other path, characterized by myopic competition, inequity, and a lack of oversight, leads toward a dystopia of destabilizing weapons, entrenched hierarchies, and existential technological risks.

The theorists from Chapter 1 provide our compass. Feynman gave us the vision of the possible. Drexler provided the engineering roadmap and the stark warnings. Blumer offers the sociological lens to understand the tumultuous transition. The literature reviewed in this chapter confirms that both their promises and their warnings are already beginning to manifest. The trajectory is not predetermined. It is the collective project of our time to steer the confluence toward a future that reflects our highest aspirations, not our deepest fears. The following chapters will delve into the specific domains of policy, economics, and ethics where this steering must occur.

CHAPTER 3: CURRENT ADVANCES IN NANOTECHNOLOGY: A DUAL-EDGED SWORD

Introduction

The trajectory of nanotechnology, from Drexler's seminal conceptualization to its current status as a cornerstone of 21st-century innovation, validates its characterization as a profoundly dualistic force. As elucidated in Chapter 2, nanotechnology holds the simultaneous promise of societal renaissance and existential risk—a dichotomy framed by the theories of Feynman, Drexler, and Blumer. This chapter synthesizes contemporary research to analyze the current state of advances across key domains, deliberately focusing on their inherent mixed impacts. Each technological stride forward is accompanied by a parallel shadow of ethical, environmental, or socio-economic concern, underscoring the critical need for anticipatory governance and nuanced public discourse.

3.1 Global Policy Initiatives: Convergence and Sovereignty

The transnational nature of nanoscale science has catalyzed unprecedented global policy initiatives aimed at harmonizing development, safety, and commerce. The U.S. National Nanotechnology Initiative (NNI) and its 2021 Strategic Plan exemplify a concerted effort to maintain competitiveness while ostensibly addressing ethical and safety concerns through inter-agency coordination and international engagement. Similar frameworks exist in the EU, Japan, and China, promoting data sharing and collaborative risk research.

Positive Impact: These initiatives foster international scientific collaboration, accelerate innovation, and aim to establish common safety protocols (e.g., the tiered testing approach recommended by Balbus et al., 2007). They seek to prevent a fragmented regulatory landscape that could hinder trade and safety. Platforms like the "Nano Scenario" simulation (Jarmon et al., 2008) represent innovative efforts to democratize policy discourse, enhancing public literacy and enabling "effective citizenry participation."

Negative Impact: Global policy forums are often dominated by technologically advanced nations and corporate interests, potentially marginalizing the Global South in standard-setting processes. As Hede (2007) warns, this could institutionalize a "societal and organizational divide into poor and rich, weak and strong." Furthermore, the emphasis on competitiveness and innovation can sometimes overshadow the precautionary principle, leading to a policy environment that prioritizes market entry over comprehensive risk mitigation. The tension between proprietary intellectual property (a driver of innovation) and the open collaboration needed for safety testing remains a significant hurdle.

3.2 New, Reliable, and Cost-Effective Solutions: Market Revolution and Disruption

Advances in nanomanufacturing, such as the FILM process for tissue engineering (Fozdar et al., 2008), and the development of nanocomposites using nanoclays (Uddin, 2008), are yielding materials with exceptional strength, reactivity, and novel functionalities. These materials promise longer-lasting, more efficient, and cheaper products, from lightweight automotive parts to advanced textiles.

Positive Impact: The economic and practical benefits are immense. Reliable, high-performance materials can reduce waste and energy consumption. Cost-effective production, potentially approaching the vision of molecular assemblers, could dramatically lower the price of essential goods, raising the global standard of living. Uddin (2008) documents the rising demand for nanoclay-polymer composites, highlighting their commercial viability.

Negative Impact: This rapid material revolution poses severe market disruption. Traditional industries and supply chains may collapse, leading to significant job displacement. Palmberg's (2008) analysis of the "valley of death" between academia and industry in Finland hints at the broader economic turbulence that can accompany technological transition. Furthermore, the drive for cost-effectiveness can incentivize production in regions with lax environmental and labor regulations, exporting risk. The longevity and novel chemical properties of nanomaterials also complicate end-of-life recycling, potentially creating a new legacy of persistent waste.

3.3 Intelligent Products and Nanomedicine: Precision and Peril

The convergence of AI and nanotechnology is manifesting in intelligent, responsive systems. In medicine, this takes the form of targeted therapeutic nanoparticles, like those for multi-drug-resistant cancer (Yadav et al., 2009), and advanced tissue scaffolds.

Positive Impact: The advances are revolutionary. Smart drug delivery systems promise to silence disease mechanisms (like MDR-1 genes) with minimal side effects, moving from treatment to cure. Diagnostic nanobots, as speculated by Kurzweil (cited in Kennedy, 2009), represent a frontier of personalized, preventative medicine. These advances directly combat human suffering and could significantly extend healthy lifespans.

Negative Impact: These same capabilities raise alarming ethical and safety questions. Intelligent, pervasive nanoscale devices threaten personal privacy and autonomy. In medicine, the high cost of developing nano-therapies could exacerbate healthcare inequities, creating a "nano-divide" where only the wealthy access enhancement and life-extension treatments. The long-term biological fate and toxicity of sophisticated nanoparticles remain poorly understood (Egbuna et al., 2021). Their potential for dual-use is stark; a technology designed to repair tissue could be weaponized to disrupt physiological functions.

3.4 Advances in Science and Medicine: The Frontier of Self-Modification

Nanotechnology is not merely a tool for treating disease but is becoming a platform for human enhancement. Research into artificial blood cells ("respirocytes") and neural interfaces points toward a future where human biology is actively augmented.

Positive Impact: Beyond curing illness, nanotechnology could enhance cognitive and physical performance, repair traumatic injuries completely, and potentially offset aging processes. This represents a fundamental leap in human agency over our own biology.

Negative Impact: This path leads to profound ethical quandaries. It challenges definitions of *normal* and *human*, potentially stigmatizing those who are unenhanced. It could create unprecedented social stratification based on access to enhancement. Furthermore, as Horner (2005) skeptically notes, the unintended, emergent consequences of such profound self-modification are impossible to predict, risking unforeseen biological or social pathologies.

3.5 Impacts on Environment, Health, and Safety: The Rise of Nanotoxicology

The field of nanotoxicology has emerged directly in response to the mixed impacts of nanotechnology. Studies like Benn and Westerhoff's (2008) on nanosilver leaching from textiles provide concrete data on environmental pathways, showing how nanoparticles evade traditional water treatment and accumulate in biosolids.

Positive Impact: Proactive nanotoxicology research enables a safer-by-design approach. The interdisciplinary workshop reported by Balbus et al. (2007) represents the scientific community's attempt to establish rigorous hazard assessment protocols alongside development. This knowledge allows for the creation of regulatory frameworks and safer handling procedures (Schulte & Salamanca-Buentello, 2007).

Negative Impact: The current reality is one of significant risk and uncertainty. The unique properties of nanoparticles—high reactivity, mobility, and ability to cross biological barriers—make them potent toxicants. Inhalation of certain nanoparticles can cause pulmonary inflammation akin to asbestos (Oberdörster et al., 2005). Their ecotoxicological effects are concerning; they can disrupt microbial communities essential for ecosystem health. A persistent gap exists between the rapid commercialization of nano-enabled products and the slow pace of definitive risk assessment, leaving workers, consumers, and ecosystems in a de facto experiment.

3.6 The Risk Perception Chasm: Experts, Public, and Policy

A critical social impact lies in the divergence of risk perception, as explored by Powell (2007) and Anderson et al. (2007). "Upstream" scientists and developers often perceive risks as manageable and distant, while "downstream" toxicologists and the public view them as substantial and immediate.

Impact: This chasm threatens social license. If public trust is eroded—a lesson from the GM crops debate—backlash can stifle beneficial innovation. It can also lead to poor policy, as fragmented scientific standpoints (Powell, 2007) provide unclear guidance to regulators. Effective communication and transparent, inclusive deliberation are not peripheral concerns but central to the sustainable development of nanotechnology.

Conclusion

The current advances in nanotechnology vividly embody Drexler's vision of a technology with boundless potential for creation and destruction. Each category of progress—from global policies and smart materials to medical nanobots—generates a corresponding spectrum of consequences. The emergence of nanotoxicology as a formal discipline is a telling indicator that the negative impacts are not speculative but present and demanding of rigorous science.

The societal impact of nanotechnology is therefore not a future event to be predicted, but an ongoing process to be managed. It is shaped by the constant interplay between innovation and risk, between economic promise and ethical peril, and between expert assessment and public trust.

Navigating this complex landscape requires moving beyond simplistic narratives of hype or alarm (Stang & Sheremeta, 2006) and embracing a nuanced, interdisciplinary, and precautionary approach that seeks to steer the immense power of nanotechnology toward equitable human and environmental flourishing. The next chapter will synthesize these mixed impacts into a cohesive framework for understanding and guiding the social change inherent in the nanotechnological revolution.

CHAPTER 4: THE CONVERGENT HORIZON – ALLIED TECHNOLOGIES SHAPING TOMORROW

Introduction: From Singular Trends to a Compound Future

The preceding chapter established that emerging technologies like nanotechnology are not developed in isolation; they are social agents with inherently dualistic impacts. As we look forward, it becomes clear that the future will not be defined by the linear progression of a single technology, but by the convergence and compounding of multiple technological waves (McKinsey & Company, 2025). This chapter examines the trajectory of key allied technologies—Robotic Process Automation (RPA), Virtual and Augmented Reality (VR/AR), Blockchain, the Internet of Things (IoT), and Driverless systems. It will analyze how these trends are moving beyond their initial domains to intersect with each other and with foundational forces like Artificial Intelligence (AI), creating new capabilities, reshaping industries, and presenting novel societal challenges.

The central thesis of this analysis is that the most significant advances and disruptions will occur at the intersections of these technologies. Intelligence is moving from the cloud to the physical edge, autonomy is scaling from digital tasks to physical systems, and trust is being engineered into decentralized networks. This convergence demands a holistic understanding, as the societal and ethical frameworks discussed in earlier chapters must now grapple with a technological landscape of unprecedented complexity and interdependence.

4.1 Robotic Process Automation (RPA): Evolving into Agentic Automation

Initially conceived as software "robots" for automating repetitive, rules-based digital tasks, RPA is undergoing a fundamental transformation. It is evolving from task-specific automation into what leading analysts now term "agentic AI" or a "silicon-based workforce" (Deloitte Insights, 2025).

Current State and Trajectory: Traditional RPA excelled at structured data entry, form processing, and basic workflow management. The trend, however, is toward intelligent agents that can autonomously plan and execute multi-step workflows, make context-aware decisions, and learn from outcomes (Deloitte Insights, 2025). This shift is not merely about faster software; it represents the creation of virtual coworkers capable of managing complex processes end-to-end. Intelligence is becoming embodied, autonomous, and solving real problems in various industries and fields of application. For example, platforms like Microsoft Copilot for Microsoft 365 and UiPath Autopilot exemplify this transition. They act as AI-powered agents that can, for instance, autonomously process a vendor invoice by extracting data, cross-referencing it with procurement records and contract terms, resolving discrepancies by querying a database, and initiating payment—all while logging its decisions for human review. This moves automation from mimicking clicks to executing judgment-driven processes.

Convergence with AI and Analytics: This evolution is powered by the fusion of RPA with advanced AI, machine learning (ML), and predictive analytics. Modern RPA platforms integrate natural language processing (NLP) to interpret unstructured documents and computer vision to "see" and interact with on-screen information. They leverage ML to identify optimization opportunities within processes themselves. Critical AI Role: Large Language Models (LLMs) are the catalyst, enabling these agents to understand natural language instructions, interpret ambiguous data, and generate human-like reasoning for their actions. For instance, an agentic AI can read a complex customer service email, understand the emotional tone and multiple issues presented, access several backend systems to compile a history, and draft a comprehensive, personalized response for a human agent to approve (Deloitte Insights, 2025).

Societal and Economic Impact: The implications are profound. While RPA was often seen as a tool for back-office efficiency, agentic automation threatens to reshape knowledge work in legal review, financial analysis, and mid-level management. Research indicates that while 38% of organizations are piloting agentic solutions, only 11% have them in production, highlighting a significant implementation gap (Deloitte Insights, 2025). The risk, as experts warn, is not that the technology fails, but that organizations fail to redesign processes for this new capability. The societal impact mirrors that of AI-driven job displacement, but with a focus on procedural and administrative professions (Business Insider, 2024; Vallance, 2023). Success requires a strategic shift from labor replacement to operational redesign, creating new roles in AI agent oversight, process transformation, and digital workforce management.

4.2 Virtual Reality (VR) and Augmented Reality (AR): From Immersion to Integration

VR and AR are transitioning from niche applications in gaming and entertainment toward becoming integral tools for enterprise, collaboration, and real-world interaction. The trend is away from standalone experiences and toward seamless integration into professional workflows and daily life.

Hardware and Experience Evolution: The hardware is becoming more sophisticated, with trends pointing toward headsets offering higher-resolution displays, increased sensory feedback, and wireless connectivity (TechJury, 2024). More importantly, the applications are maturing. Companies like Microsoft (with HoloLens 2) and Magic Leap are deploying AR in manufacturing for complex assembly guidance, where holographic arrows and instructions are overlaid directly onto engine blocks, reducing errors and training time by up to 30%. In medicine, platforms like Proximie use AR to allow expert surgeons to virtually "scrub in" and guide remote procedures by drawing annotations directly onto the local surgeon's field of view.

Convergence with AI and IoT (Spatial Computing): The future of VR/AR lies in its convergence with AI and IoT to create contextual, intelligent environments—often called *spatial computing*. AI computer vision enables devices to understand and map the 3D physical environment in real-time, allowing virtual objects to interact realistically with surfaces and occlusions. Generative AI can create dynamic, interactive virtual assets on the fly. IoT integration allows AR devices to visualize and interact with data streams from connected sensors and devices. For instance, a technician wearing AR glasses at a factory could see not just a static manual, but an AI-generated, animated overlay showing the optimal disassembly path for a specific pump model, while IoT data streams show real-time pressure readings and thermal hotspots overlaid on the physical unit.

Societal and Ethical Impact: As these technologies become more pervasive, they raise significant questions. Issues of data privacy, digital addiction, and physical safety (e.g., distraction in real-world environments) are paramount (Koenig Solutions, 2023). Furthermore, the ability of AR to persistently collect visual and behavioral data from a user's surroundings creates unprecedented surveillance potential. Current Concern: The development of AI-powered *smart glasses* for consumers raises fears of continuous, passive facial recognition and environmental recording. There is also a risk of a *digital divide* in access to these transformative tools for education and work. The ethical development of VR/AR must prioritize user well-being, establish clear boundaries for data collection, and ensure these technologies augment human capability without eroding physical-world social connections.

4.3 Blockchain: Expanding Beyond Cryptocurrency

Blockchain technology is steadily progressing from its foundational association with cryptocurrencies toward providing trust, transparency, and efficiency in complex, multi-party systems. The trend is toward application-specific platforms and integration with other technologies to solve tangible business problems.

Trends in Application and Scalability: The focus has shifted from general-purpose chains to specialized blockchain solutions for sectors like supply chain, healthcare, and digital identity. For Example: IBM Food Trust uses blockchain to trace food products from farm to shelf. Partners like Walmart can trace the origin of mangoes in seconds versus days, dramatically improving food safety response. In trade finance, platforms backed by major banks use blockchain to automate and secure letters of credit, reducing processing from 5-10 days to under 24 hours. These applications leverage blockchain's core strengths—immutability, provenance tracking, and decentralized consensus—to combat fraud, ensure authenticity, and streamline audits. However, challenges of scalability, energy consumption (for some consensus mechanisms), and interoperability between different blockchains remain active areas of innovation.

Convergence with IoT and AI: Blockchain finds powerful synergies with IoT and AI. For IoT, blockchain can provide a secure, tamper-proof ledger for the massive volumes of data generated by devices. Here, AI acts as the analytical layer. Smart contracts on a blockchain can be programmed to execute automatically based on IoT data (e.g., release payment when a shipment's temperature sensor confirms safe arrival). AI models can analyze the immutable IoT data stored on-chain to predict maintenance needs or optimize logistics, with the blockchain providing a verifiable audit trail of the data used for each AI decision, helping to address the *black box* problem and manage data provenance used in training models. This convergence is critical for building trusted autonomous systems, such as in decentralized energy grids where AI optimizes distribution and blockchain settles peer-to-peer transactions between solar panels and consumers.

Societal and Regulatory Impact: The societal promise of blockchain lies in its potential to disintermediate institutions and return control of data and assets to individuals. This has implications for voting systems, personal medical records, and intellectual property. However, it also poses regulatory challenges in areas like legal enforcement, taxation, and consumer protection in decentralized environments. The tension between the technology's libertarian origins and the need for governance frameworks will shape its adoption. Additionally, the environmental impact of certain blockchain implementations continues to spur innovation toward more sustainable consensus models like Proof-of-Stake (PoS).

PoS is a blockchain mechanism designed to validate transactions and secure a network without the enormous energy consumption associated with earlier models like Proof-of-Work (PoW). In a PoS system, the right to validate transactions and create new blocks are given to participants who "stake" their cryptocurrency—locking up a portion of their coins as collateral.

4.4 Internet of Things (IoT) and the Intelligence at the Edge

IoT is evolving from a network of connected consumer devices to the central nervous system of smart industries, cities, and infrastructure. The trend is toward more intelligent, secure, and autonomous edge devices that process data locally.

From Connectivity to Predictive Action: The initial wave of IoT was about connectivity and data collection. The current trend, powered by cheaper sensors and better analytics, is toward predictive analytics and autonomous response. For example: John Deere has transformed its equipment into AI-driven *smart farms on wheels*. Tractors equipped with IoT sensors and computer vision cameras analyze soil conditions and crop health in real-time. Onboard AI models make micro-decisions, adjusting seed depth and fertilizer application on a per-plant basis, optimizing yield and reducing chemical use by significant percentages. Systems no longer just monitor temperature or vibration; they predict equipment failure and schedule maintenance, or optimize energy usage across a smart grid in real-time. The World Economic Forum notes that faster 5G and IoT connections could unlock significant economic activity, potentially increasing global GDP by trillions by 2030 through applications in mobility, healthcare, and manufacturing (McKinsey & Company, 2025).

Convergence with Edge Computing and AI: This predictive capability comes from the convergence of IoT, edge computing, and AI. Deploying lightweight AI models directly on IoT devices has become standard practice, with platforms like NVIDIA Jetson enabling edge-based robotics and video analytics. In manufacturing, Siemens and Intel use edge AI to analyze motor vibration data in real time, predicting failures days in advance and automatically triggering work orders—cutting latency from minutes to milliseconds and reducing bandwidth use (McKinsey & Company, 2025). Edge computing processes data at or near the IoT device, enabling real-time decisions and minimizing cloud dependence—critical for applications such as autonomous vehicles, industrial robotics, security cameras, and assembly-line anomaly detection (McKinsey & Company, 2025; Koenig Solutions, 2023). Combined with lightweight AI models, this creates a distributed and resilient network of intelligent endpoints.

Societal, Security, and Environmental Impact: The scaling of IoT presents a massive expansion of the digital attack surface, making cybersecurity a paramount concern (Koenig Solutions, 2023). The 2016 Mirai botnet attack, which hijacked millions of insecure IoT cameras, underscores this risk. The proliferation of devices also raises serious issues of data privacy, as intimate details of personal lives and organizational operations are continuously sensed and transmitted. From an environmental perspective, IoT is a dual-edged sword: it enables dramatic efficiencies in resource use (smart agriculture, energy grids) but also contributes to electronic waste and energy consumption from billions of always-on devices. Managing this impact requires robust security-by-design principles, strong data governance, and responsible lifecycle management for IoT hardware.

4.5 Driverless Technology and the Autonomous Ecosystem

Driverless technology represents one of the most complex convergences of allied technologies, aiming to create safe, reliable, and scalable autonomous transportation systems. Its progression is tied directly to advances in nearly every other trend discussed.

Core Technological Pillars: The functionality of driverless systems rests on a triad of technologies:

> Advanced Sensors and IoT: Networks of LiDAR, radar, cameras, and ultrasonic sensors act as the vehicle's eyes and ears, generating massive, real-time data streams about the environment.
> AI and Machine Learning: This is the system's brain. Companies like Waymo and Cruise use deep learning for *perception* (identifying and classifying objects) and reinforcement learning for

planning (making navigation decisions). Their AI is trained on billions of miles of simulated and real-world driving data to handle rare *edge cases* (e.g., a ball rolling into the street followed by a child).

Edge Computing and Connectivity: Onboard computing hardware (edge computing) processes data instantaneously for safe vehicle control. Simultaneously, vehicle-to-everything (V2X) communication, a form of IoT, allows cars to share information with each other and with smart city infrastructure, enhancing collective awareness.

Trends Beyond Personal Cars: While consumer autonomous vehicles capture public imagination, significant near-term progress is occurring in constrained environments. This includes autonomous logistics in warehouses (e.g., Amazon's deployment of over a million robots coordinated by AI) and "self-driving" systems within industrial factories, such as BMW's use of autonomous carriers in production (Brode, 2022). These applications provide controlled, valuable proving grounds for the technology. For example: Waymo Via is operating autonomous Class 8 trucks on freight routes in Texas. In mining, companies like Rio Tinto use fully autonomous haul trucks in Australian mines, operated by centralized AI systems that optimize routes and fuel efficiency 24/7, showcasing the maturity of the technology in controlled settings.

Societal, Ethical, and Urban Impact: The societal implications are vast. Promised benefits include reduced traffic fatalities (eliminating human error), increased mobility for non-drivers, and optimized traffic flow. However, they introduce profound ethical dilemmas (algorithmic decision-making in accident scenarios), legal and liability challenges, and potential for significant workforce displacement in transportation and logistics sectors (Business Insider, 2024). Besides, the widespread adoption of driverless technology could radically reshape urban design (less need for parking, repurposed roads), land use, and the very concept of vehicle ownership, demanding proactive policy and civic planning.

Conclusion: Navigating the Convergent Landscape

The trajectory of these allied technologies underscores a fundamental shift: the era of siloed technological advancement is over. The future belongs to convergent systems where AI powers autonomous agents, blockchain provides trust for IoT data, and edge intelligence enables real-time AR interactions and driverless navigation.

This convergence amplifies both the potential benefits and the inherent risks. It promises unprecedented efficiency, personalization, and solutions to global challenges in healthcare, sustainability, and logistics. Yet, it also concentrates complexity, creates interconnected points of failure, and raises synergistic ethical concerns—from biased autonomous systems making physical-world decisions to pervasive surveillance networks fueled by IoT and AI.

Therefore, the central challenge for researchers, policymakers, and society—building upon the frameworks of Drexler, Blumer, and others—is to steer this convergence deliberately. It requires interdisciplinary collaboration, anticipatory governance that focuses on systems rather than isolated technologies, and a continued commitment to prioritizing human dignity, equity, and safety in the design and deployment of the compound technological landscape that defines our coming decade.

CHAPTER 5: PREDICTIVE SOCIAL IMPACTS OF NANOTECHNOLOGY AND ARTIFICIAL INTELLIGENCE

Introduction

The convergence of allied technologies, as detailed in Chapter 4, forms the critical infrastructure upon which advanced nanotechnology will be built and scaled. The trends in intelligent automation, ubiquitous sensing, and autonomous systems are not separate from nanotechnology; they are its enabling partners and primary channels for societal impact. This chapter synthesizes the predictive social outcomes of nanotechnology, moving from its material foundations to its ultimate implications for human life, governance, and the global order. Building upon current research trajectories, we present a holistic analysis of nanotechnology's dualistic potential to reshape civilization, now actively accelerated and mediated by Artificial Intelligence.

5.1 New Products and Material Foundations

The primary vector for nanotechnology's societal entry will be through novel products, born from revolutionary material classes engineered and optimized with AI.

Nanomaterials and Nanoelectronics: The development of materials like carbon nanotubes, graphene, and quantum dots enables a fundamental leap in product capabilities. AI and machine learning are now indispensable in this domain. Researchers at institutions like MIT and Stanford use generative AI models to predict the properties of never-before-synthesized nanomaterials, dramatically accelerating discovery (Brode, 2022). For example, AI can screen millions of potential molecular structures for a new battery electrolyte that is stable, conductive, and inexpensive, a process that would take humans centuries. This leads to nanoelectronics that are not merely smaller but qualitatively different—enabling flexible, biocompatible sensors and ultra-low-power processors for the pervasive IoT networks discussed in Chapter 4.

Nanocatalysts and Clean Production: Nanocatalysts promise radical efficiency in industrial processes like chemical synthesis and energy conversion. Companies like Citrine Informatics use AI platforms to analyze materials data and guide the development of nanocatalysts that reduce energy consumption in manufacturing by up to 50%. The predictive impact is the enablement of circular economies, where AI-designed nanocatalysts break down plastic waste at the molecular level for repolymerization, directly addressing sustainability goals.

Reliable and Intelligent Products: The inherent precision of atomic-scale manufacturing theoretically points to products of exceptional reliability. AI Convergence: Here, AI provides the "mind" for the nanotech "body." A product with embedded nanosensors can monitor its own structural integrity. AI algorithms process this data stream to predict failure and even initiate self-repair by triggering a response from encapsulated nano-agents. This creates a new paradigm of "living" products that maintain themselves, reducing waste and transforming ownership models.

5.2 Transformations in Health and Demography

Nanotechnology's most profound personal impacts will be felt in medicine and human longevity, with AI acting as the essential control and diagnostic system.

Nanomedicine and Improved Science: Research into targeted drug delivery, hemostatic agents, and imaging probes represents the vanguard of nanomedicine. This directly leads to improved science and inventions, as nanotechnology provides new tools to understand and intervene in biological systems at the molecular level. AI-

Integrated Example: The company Berg uses AI to analyze biological data (proteomics, metabolomics) to identify novel drug targets, then designs nanocarriers to deliver therapies precisely to diseased cells. In diagnostics, Google Health is developing AI algorithms that work with nanoparticle-based contrast agents to detect cancers like metastatic breast cancer in MRI scans years earlier than current methods. This synergy leads to a shift from generalized treatment to predictive, personalized healthcare.

Longevity and Population Crises: The logical extension of advanced nanomedicine—including concepts like reparative nanobots—is a significant extension of the human healthspan. AI's Critical Function: Managing such complex in vivo systems is impossible without sophisticated AI. The vision of *nanosurgeons* would require distributed AI to coordinate billions of nanodevices, monitor vitals, and make real-time therapeutic decisions. This forces a direct confrontation with the anticipated challenges of longevity and population pressures. If successful, societal systems for retirement, healthcare, resource allocation, and even family structures would require fundamental redesign to accommodate dramatically extended lifespans and potential population growth. The resulting strain on pension systems and the challenge to intergenerational equity would necessitate a comprehensive rethinking of social contracts.

5.3 Security, Governance, and Control

The power of nanotechnology will inevitably be applied to the domains of security and authority, creating new paradigms of control and conflict that are automated and enhanced by AI.

Crime Fighting and Governing: Enhanced surveillance through nano-sensors (e.g., dust-sized airborne sensors) and ubiquitous tracking enables predictions of superior crime-fighting capabilities. AI is the force multiplier. It would analyze feeds from trillions of nanosensors, using facial recognition, gait analysis, and anomaly detection to identify threats in real-time. China's social credit system, combined with extensive surveillance, provides a precursor to this AI-nano governance model. This creates a tension between security and privacy, potentially leading to automated policing systems with minimal human oversight, raising profound questions about algorithmic bias and due process.

Warfare and Mass Destruction: Military research into nano-enhanced materials (e.g., lightweight, super-strong armor) and sensors is active. The AI-Nano Weapon Convergence: The most disruptive (and dystopian) prediction involves AI-controlled nano-swarms. These could be programmable materials that disable electronics, deliver targeted biochemical agents, or perform mechanical sabotage. The U.S. Defense Advanced Research Projects Agency (DARPA) has programs investigating *organized matter*. When governed by an AI, such a swarm could be an agent of mass destruction that is scalable, precise, and deniable, lowering thresholds for conflict and creating a new, terrifying class of weaponry that challenges traditional deterrence models.

The combination of AI-enabled nanoscale hardware and facial recognition technology raises serious ethical and security concerns, particularly regarding the potential misuse of autonomous systems to deliver harmful agents to specific individuals. Scholars and security analysts have warned that advances in miniaturized drones and AI-driven targeting could enable highly precise, automated attacks. For example, the 2018 U.S. National Academies of Sciences report on "Reducing the Threat of Chemical, Biological, Radiological, and Nuclear Terrorism" noted that AI-assisted autonomous systems could, in theory, be adapted for targeted delivery of toxic substances, underscoring the need for strong governance and safeguards (National Academies of Sciences, Engineering, and Medicine, 2018).

5.4 Quality of Life and Socioeconomic Reconfiguration

The aggregate effect of the above impacts will trigger a deep reordering of daily life and economic structures.

Environmental Impact and Use of Time: Nanotechnology offers tools for environmental remediation (e.g., water purification, nano-sponges for oil spills, photocatalytic nanoparticles for air purification) but also poses novel risks of nano-pollution and toxicity. AI for Safety and Management: AI is crucial for modeling the long-term environmental fate of nanomaterials and managing their lifecycle. Furthermore, if molecular manufacturing ever matures, it could automate production of physical goods to a radical degree. The predicted problem of "use of time" emerges. In such a post-scarcity economy, AI would manage production and resource allocation, while human purpose decouples from traditional work. This would require new cultural, educational, and social models to address potential crises of meaning and mental health.

The Risk of Nano-Colonization: The high barrier to entry for advanced nanotechnology, compounded by the compute power and data needed for its AI drivers, could exacerbate global inequalities. Current Precedent: The global

AI chip shortage and concentration of talent in a few tech hubs illustrate this dynamic. The predictive social impact is a new form of geopolitical stratification—a nano-colonization where technologically dominant nations or corporations control the essential means of production, medicine, and defense. This could render traditional economic and political leverage obsolete for non-nano powers, creating a world of deepened dependency and control, potentially locking in a permanent global hierarchy.

Conclusion: Moral and Ethical Considerations in a Nano-Scale Future

The predictive social impacts outlined above are not deterministic; they are contingent on choices made today. The trajectory of nanotechnology, deeply intertwined with AI, will be decided in the interplay between technical feasibility and human values. Several core moral and ethical considerations must anchor this development:

> *Equity and Justice:* The convergence of nano and AI could be the greatest force for inequality in history. Global policy initiatives, like the UN Advisory Body on AI, must expand to include nanotechnology, prioritizing equitable access frameworks over proprietary exclusivity.

Autonomy and Privacy: In a world of intelligent nano-surveillance, preserving cognitive liberty and private thought is paramount. Ethical frameworks must establish technological bright lines, such as bans on pervasive neural monitoring or environmental nano-surveillance.

Safety and the Precautionary Principle: Given the potential for irreversible environmental damage and novel weapons, a robust precautionary principle must be integrated into the innovation cycle. Research into nanotoxicology, led by AI simulation, is a fundamental ethical requirement (John et al., 2022).

Control and Responsibility: As AI-controlled nano-systems become possible, the chain of accountability must be legally and technically explicit. The principle of "meaningful human control" must be architected into any autonomous physical system, especially at the nanoscale.

The Redefinition of Humanity: Technologies that promise radical longevity, enhanced cognition, and intimate human-machine integration challenge our very definition of what it means to be human. A broad, inclusive, and profound societal dialogue is needed to navigate this future consciously.

In summary, nanotechnology—amplified and accelerated by its convergence with AI and other strategic technologies—stands as the quintessential social change agent of the 21st century. Its predictive impacts span from the miracle of personalized disease eradication to the terror of new mass destruction, from the promise of material abundance to the peril of a new global hierarchy. The destination depends on our collective wisdom, foresight, and ethical courage to steer this immense, compound power toward the ennoblement, rather than the erosion, of the human future.

PART II

HANDBOOK FOR OPERATORS OF CONSUMER GOODS MANUFACTURING COMPANIES IN THE AGE OF AI

Introduction

The global consumer goods market is in a state of unprecedented flux. Driven by accelerating e-commerce, direct-to-consumer models, and hyper-personalization, the very definition of supply chain efficiency and customer satisfaction is being rewritten. At the heart of this transformation lies a powerful, disruptive force: Artificial Intelligence (AI). For operators of manufacturing companies—especially those in the high-velocity, low-margin contract manufacturing niche—this presents both an existential challenge and a once-in-a-generation opportunity.

This handbook is not a speculative look into a distant future. It is a practical, operations-focused manual for the here and now. Its purpose is to demystify the application of AI within the factory walls and executive offices of consumer goods manufacturing. We move beyond theoretical discussions to provide actionable strategies, frameworks, and considerations for integrating intelligent systems into every facet of your operation: from predictive maintenance on the production line to hyper-accurate demand forecasting for your brand partners.

The audience for this book is the operational backbone of the industry: the Plant Managers, Operations Directors, Supply Chain VPs, and business owners who must deliver quality products on time and at cost, every single day. You are tasked with navigating the complexities of contract manufacturing—managing volatile demand, ensuring rigorous compliance, and maintaining razor-thin margins—all while the technological ground shifts beneath your feet. This book is your guide to not just surviving this shift, but leveraging it to build a more resilient, agile, and competitive enterprise.

We will systematically explore how AI transforms core operational pillars: production planning, quality control, supply chain optimization, and human resource management. Each chapter blends strategic insight with practical implementation steps, featuring real-world case studies of early adopters. We will also confront the critical challenges head-on: the ethics of automation, the imperative of workforce reskilling, data security in interconnected systems, and calculating the tangible ROI of AI investments.

The convergence of physical manufacturing and digital intelligence is creating the "Smart Factory" and the "Cognitive Supply Chain." This handbook is your essential roadmap to navigating this new landscape. By embracing the principles and practices within, you can transform AI from a buzzword into your most powerful tool for operational excellence and sustainable growth.

FOUNDATIONS FOR A NEW ERA

CHAPTER 1: THE AI-INFUSED LANDSCAPE OF CONSUMER GOODS MANUFACTURING: FOUNDATIONS FOR A TRANSFORMATIVE ERA

Introduction: A Paradigm Shift in Production

We stand at the inflection point of a new industrial age, often termed Industry 4.0, characterized by the cyber-physical integration of manufacturing systems (Kagermann et al., 2013). The consumer goods manufacturing sector, long characterized by economies of scale and predictable linear supply chains, is undergoing a profound and permanent transformation. This change is not driven by a single invention but by the convergence of digital intelligence with physical production—a fusion creating what is now termed the "cognitive factory" and the "self-optimizing supply chain" (Zhou et al., 2019).

This chapter serves as the foundational map for this new terrain. It diagnoses the relentless external pressures reshaping the industry's very purpose and then demystifies the core technological arsenal—Artificial Intelligence (AI)—that will define competitive survival. Finally, it reimagines the human role at the center of this intelligent system. For operators, engineers, and managers, understanding this landscape is no longer an academic exercise; it is the critical first step in a necessary journey from reactive executors of a plan to proactive architects of a resilient, agile, and intelligent production ecosystem. The era of managing machines is giving way to the era of managing intelligence.

1.1 The New Imperatives: Redefining the Rules of Competition

The traditional manufacturing playbook, optimized for cost and volume, is obsolete. Three interconnected imperatives—Speed, Customization, and Sustainability—now form the triple constraint within which all successful operations must be designed. These are not niche trends but fundamental shifts in consumer expectation and regulatory reality, directly transmitted to contract manufacturers through their brand partners.

1.1.1 The Amazon Effect: The Compression of Time and the Death of Inventory Buffers

The "Amazon Effect" transcends a single retailer; it represents a seismic reset of the customer's psychological clock. It fundamentally rewires customer expectations, creating a new baseline for what feels normal or timely. The anticipation of near-instantaneous, free, and perfectly tracked delivery—whether B2B or B2C—has irrevocably altered the calculus of production planning (Huang & Li, 2022). For contract manufacturers, this represents a paradigm shift and a redefinition of agility, rendering all other sub-standards undesirable and obsolete.

From Forecast-Driven to Demand-Responsive: The classic model of producing large batches against a quarterly forecast, storing finished goods, and fulfilling from warehouse stock is financially perilous. It leads to endemic issues: dead stock of unsold items and stockouts of trending ones. AI-powered *demand sensing* replaces this with a dynamic model. By analyzing real-time data streams—point-of-sale data, social media trends, search engine queries, and even local weather—AI models can detect demand signals weeks earlier than traditional methods (Choi et al., 2021). For the manufacturer, this means shifting production schedules from a monthly to a weekly or even daily rhythm, aligning output precisely with consumption.

The Redesign of Inventory Strategy: The old safety-stock paradigm is a capital trap. The new imperative is *inventory velocity*. AI enables a shift to a *just-in-sequence* model where raw materials arrive and products are finished just in time for immediate shipment, often in mixed-SKU pallets configured for direct-to-consumer or store-specific delivery. This requires exquisite synchronization with suppliers and logistics partners, orchestrated by AI-driven supply chain control towers that provide end-to-end visibility and proactive exception management.

Operational Agility as a Core Competency: The ability to execute a rapid changeover is no longer a nice-to-have; it is a survival skill. A contract packager, for instance, must be able to switch from producing 10,000 units of a health drink for Retailer A to 5,000 units of a premium skincare product for Brand B within hours, not days. AI and robotics enable this through *digital work instructions* that automatically reconfigure lines, and smart tools that guide technicians, minimizing downtime and human error.

1.1.2 Mass Customization: The Efficiency Paradox and Its AI Resolution

The drive for personalization represents perhaps the most complex operational challenge: reconciling infinite variety with industrial efficiency. Consumers now expect products tailored to their preferences—be it a nutritionally personalized vitamin blend, sneakers with custom colorways, or cosmetics matched to their skin tone. For manufacturers, this means moving from batches of millions to lots of one, without eroding margins.

The Challenge of Immense Variability: Each variant introduces complexity: unique bills of materials (BOMs), specific assembly instructions, distinct packaging, and customized quality control checks. Traditional Enterprise Resource Planning (ERP) and Manufacturing Execution Systems (MES) strain under this complexity, leading to errors, delays, and confusion on the shop floor.

AI as the Orchestrator of Complexity: This is where AI transitions from a helpful tool to the central nervous system of the factory.

> Dynamic Scheduling: AI schedulers don't just sequence jobs; they solve a multi-dimensional optimization puzzle in real-time. They balance due dates, machine capabilities, material availability, cleaning requirements, and labor skills to create a feasible, optimal schedule for hundreds of unique orders simultaneously, re-optimizing instantly when a machine faults or a rush order arrives (Mourtzis et al., 2021).

> Guided Assembly & Error-Proofing: Augmented Reality (AR) workstations, powered by computer vision, can guide operators through custom assembly steps for each unique product. The system visually highlights which component to pick, where to

place it, and which tool to use, drastically reducing training time and assembly errors. The system can verify each step before allowing the operator to proceed, achieving *true first-pass yield* even at lot-size-one.

Flexible Automation: Collaborative robots (cobots) equipped with AI vision can be quickly reprogrammed for new tasks. One hour, a cobot is applying labels to one bottle shape; the next, it's packing a different product configuration into a mailer box. This flexibility is the physical enabler of mass customization (Matheson et al., 2022).

1.1.3 The Circular Economy & ESG: From Regulatory Burden to Operational Advantage

Environmental, Social, and Governance (ESG) criteria have moved from corporate social responsibility reports to the core of investment decisions and consumer purchasing. Regulations are tightening (e.g., Extended Producer Responsibility laws, plastic taxes), and consumers are scrutinizing the environmental footprint of products. For manufacturers, sustainability is no longer just about recycling office paper; it is a comprehensive operational mandate with direct cost implications.

AI for Material and Energy Optimization: This is where AI delivers immediate, measurable ROI while advancing sustainability goals.

Precision Consumption: In processes like injection molding, textile cutting, or food portioning, AI can analyze material properties and desired output to calculate the absolute minimum raw material required, reducing scrap by 5-15% (De Sousa Jabbour et al., 2018). In spray-coating or chemical mixing, computer vision can ensure precise application, preventing waste and reducing VOC emissions.

Predictive Energy Management: AI can model a plant's energy consumption patterns, integrating data from machines, HVAC, and lighting. It can then optimize energy use—for example, by scheduling high-energy processes for off-peak hours, pre-cooling facilities based on weather and production forecasts, or identifying inefficient equipment. This reduces both cost and carbon footprint.

Enabling the Circular Loop: AI is critical for managing products at end-of-life. Computer vision systems can rapidly and accurately sort post-consumer materials by polymer type, color, and contamination level, making recycling streams more viable and valuable (Gundupalli et al., 2017).

Furthermore, AI can aid in designing products for disassembly by simulating how different designs will fare in a recycling stream, closing the loop at the design-for-manufacturing stage.

The Sustainability Data Ledger: As brands are held accountable for Scope 3 emissions (those of their supply chain), contract manufacturers must provide auditable data on the carbon, water, and waste footprint of their production. AI systems automatically collect, calculate, and report this data from sensor networks, turning a compliance burden into a transparent competitive advantage for securing contracts with sustainability-led brands.

1.2 Demystifying AI for the Factory Floor: The Technological Toolkit

To harness these imperatives, operators must move beyond viewing AI as a monolithic, mysterious force. It is a suite of distinct, powerful technologies, each with specific applications that solve concrete operational problems.

1.2.1 Machine Learning (ML): The Engine of Prediction and Insight

At its core, ML is advanced pattern recognition. Its algorithms learn from historical data to identify patterns, correlations, and anomalies, which they then use to make predictions or decisions on new data. Its power in manufacturing lies in moving from descriptive analytics (what happened) to predictive (what will happen) and prescriptive (what should we do) insights (Wuest et al., 2016).

Predictive Maintenance: The established use case. Vibration, temperature, and acoustic sensors on a critical pump generate a continuous data stream. An ML model, trained on years of this data alongside maintenance records, learns the *signature* of healthy operation and the subtle patterns that precede a failure (e.g., a specific harmonic vibration indicating bearing wear). It can then alert technicians' weeks in advance to schedule a repair during a planned shutdown, avoiding a catastrophic, line-stopping failure and transforming maintenance from a cost center to a reliability function.

Predictive Quality: ML models can analyze process parameters (temperature, pressure, speed, raw material batch attributes) from the *beginning* of a production run and predict the quality of the *final* product. If the model predicts a high probability of a defect, the system can automatically adjust the process parameters in real-time or flag the batch for early inspection, preventing waste of time and materials.

Supply Chain Risk Forecasting: ML can ingest diverse external data—geopolitical news, port congestion reports, weather patterns, commodity futures—to model and forecast risks to raw material availability or logistics routes, allowing for proactive sourcing adjustments.

1.2.2 Computer Vision: Bestowing Industrial Sight

If ML is the brain, computer vision is the perceptive eye. It uses cameras and image-processing algorithms to extract meaningful information from visual data, performing tasks with superhuman speed, consistency, and accuracy.

Automated Visual Inspection (AVI): This is the most transformative application. AVI systems, using high-resolution cameras and deep learning models, can inspect every unit on a high-speed production line for dozens of defect types simultaneously: misprints, scratches, dents, fill-levels, cap alignment, and presence of labels. They operate 24/7 without fatigue, achieving near-100% inspection rates and providing consistent, digitized quality records (Dias et al., 2020). This not only improves quality but also frees human inspectors for more complex, value-added tasks like root-cause analysis.

Robotic Guidance and Bin Picking: Traditional robots require parts to be presented in precise, fixed positions. AI-powered vision allows robots to *see* a bin of randomly oriented parts, identify each one, and determine the optimal grasp to pick and place it accurately. This is revolutionary for assembly and packaging lines handling variable products.

Safety and Compliance Monitoring: Vision systems can monitor workspaces to ensure compliance with safety protocols (e.g., wearing protective gear in designated zones) and track the movement of goods and assets within a facility, enhancing both security and operational visibility.

1.2.3 Natural Language Processing (NLP): Bridging the Human-Machine Language Gap

NLP allows machines to understand, interpret, and generate human language. In the industrial context, it turns unstructured text and speech into structured, actionable data.

Mining the Voice of the Customer (VoC): NLP algorithms can scrape and analyze millions of online product reviews, social media posts, and customer service tickets. They can identify emerging complaints (e.g., "bottle leaks," "color fades") and quantify sentiment trends. This intelligence is fed directly back to production and quality teams, enabling them to pinpoint and rectify manufacturing issues sometimes before formal quality reports are generated.

Intelligent Documentation and Knowledge Management: NLP can read thousands of pages of maintenance manuals, standard operating procedures (SOPs), and past work orders. A technician can then query this knowledge base using natural language ("How do I troubleshoot error code E45 on the filler?") and instantly receive the relevant excerpts, diagrams, and past solutions.

Voice-Driven Workflows: In hands-busy environments, operators can use voice commands to report issues, request parts, or query system status. An operator might say, "System, report a jam at Station 3," and the AI logs the issue, notifies maintenance, and pauses the upstream line—all through speech.

1.2.4 Robotic Process Automation (RPA): The Digital Workforce for Administrative Friction

RPA is not a physical robot but a software *bot* that mimics human actions to execute repetitive, rules-based digital tasks. It is the automation of white-collar, clerical processes that plague back-office operations.

Order-to-Cash and Procure-to-Pay Automation: Bots can automatically read incoming purchase orders (PDFs, emails), enter them into the ERP system, generate pick lists, and create invoices. Similarly, they can process supplier invoices, match them to purchase orders, and initiate payments. This eliminates manual data entry errors, speeds up cycles, and reduces administrative overhead.

Compliance and Reporting: Generating regulatory reports, safety data sheets, or customs documentation is a time-consuming, error-prone process. RPA bots can be programmed to pull data from various systems, populate templates, and file reports automatically and on schedule.

Shipment Tracking and Customer Communication: Bots can monitor carrier portals for shipment status updates and automatically send tracking notifications to customers, improving the customer experience without human intervention.

1.3 The Operator's Evolving Role: From Tactical Manager to Strategic Orchestrator

The integration of these AI tools does not render the human operator obsolete; it radically redefines their value. The role shifts from direct, hands-on control of machines and reaction to alarms, to the higher-order tasks of managing, interpreting, and improving a system of intelligent processes (Romero et al., 2020).

1.3.1 The Shift in Focus: From Firefighting to System Design

The daily work moves away from:

Chasing parts and expediting shipments.

Manually adjusting machine parameters based on gut feel.

Walking the line to find quality issues.

Filling out paper checklists and reports.

And moves towards:

Monitoring AI system health and performance (Is the predictive maintenance model still accurate?).

Interpreting AI-generated insights and recommendations (The system says we should change the filter; what's the business impact if we delay?).

Orchestrating the flow of work between automated systems, robots, and human teams.

Leading continuous improvement projects informed by AI-driven data analytics.

1.3.2 The New Core Competency Set

To thrive in this new environment, operators must cultivate a blended skill set:

1. *Data Literacy and Analytical Thinking:* The ability to read dashboards, understand basic statistical outputs, and ask

the right questions of data. Operators must become comfortable with probabilities (e.g., "There's an 85% chance of failure in 14 days") rather than binary certainties.

2. *AI-Human Collaboration Management:* Knowing when to trust the AI's recommendation and when to apply human intuition and contextual knowledge. This involves understanding the limitations of models (their "edge cases") and creating effective feedback loops to retrain and improve them.

3. *Process Re-engineering Mindset:* With AI handling execution, human intelligence is freed for redesign. Operators must learn methodologies like Lean and Six Sigma, now supercharged with AI-driven data, to fundamentally reimagine workflows for greater agility, quality, and sustainability.

4. *Change Leadership and Communication:* The successful operator will be a translator and coach, explaining AI-driven changes to frontline teams, addressing concerns, and fostering a culture of experimentation and learning rather than fear of obsolescence.

Conclusion: The Inflection Point

The AI-infused landscape of consumer goods manufacturing is not a distant future scenario; it is the operational reality taking shape today. The imperatives of Speed, Customization, and Sustainability are non-negotiable market forces. The technologies of ML, Computer Vision, NLP, and RPA are the proven tools to meet them. The transformation of the Operator's Role is the necessary human evolution to wield these tools effectively.

This chapter has laid the groundwork by diagnosing the pressure and defining the tools. The following chapters will delve into the practical implementation: how to build the data foundation, pilot these technologies, scale them across operations, and navigate the critical human and ethical dimensions of this great transformation. The journey begins with a clear-eyed understanding of this new landscape, for only then can one begin to navigate its complexities and emerge as a leader in the age of intelligent manufacturing.

CHAPTER 2: BUILDING THE DIGITAL FOUNDATION: DATA, CONNECTIVITY, AND CULTURE

Introduction: The Triad of Transformation

The vision of an AI-powered, agile, and sustainable consumer goods manufacturing operation is compelling. Yet, this future cannot be built upon the brittle foundations of paper records, isolated data silos, and resistant organizational cultures. AI is not a magic wand; it is a sophisticated engine that requires *high-octane fuel*, a *robust nervous system*, and a *skilled, willing crew* to operate it. This chapter delves into the three essential, interdependent pillars that form the bedrock of any successful digital transformation: Mastering Your Data, Deploying Robust Connectivity, and Cultivating an AI-Ready Culture. Neglecting any one of these is akin to building a Formula 1 car with contaminated fuel, a faulty electrical system, and an untrained driver—a guaranteed path to failure.

This chapter is a practical guide for operators, plant managers, and IT leaders tasked with the "unsexy" but critical groundwork. We move beyond the glossy promise of AI outcomes to the gritty reality of implementation, providing a structured framework for building a digital foundation that is scalable, secure, and human-centric.

2.1 The Fuel of AI: Mastering Your Data

In the world of AI, data is not merely information; it is the fundamental raw material from which all insights, predictions, and automated actions are forged. The principle is absolute: *"Garbage In, Gospel Out."* An AI model trained on incomplete, inaccurate, or inconsistent data will produce flawed, often dangerously misleading, outputs with unwavering confidence. Therefore, data mastery is the first and most critical engineering challenge.

2.1.1 Identifying and Harvesting Data Sources: The Industrial Data Landscape

A modern manufacturing facility is a rich, often untapped, ecosystem of data streams. Building a comprehensive data strategy begins with a systematic audit to identify and instrument these sources.

Operational Technology (OT) Data: The Voice of the Machine.

Machine Logs & Controllers: This is the foundational layer. Systems like *Supervisory Control and Data Acquisition (SCADA)* and *Distributed Control Systems (DCS)* provide high-level operational data. For deeper, standardized machine-to-machine communication, protocols like *OPC Unified Architecture (OPC-UA)* are essential, enabling the extraction of real-time data on machine states, cycles, and alarms from PLCs (Programmable Logic Controllers) and CNCs (Computer Numerical Controls) across different vendors.

> IoT Sensor Networks: To move beyond what machines report, you must instrument them. A strategic deployment of IoT sensors captures the physical world's condition.
>
> Condition Monitoring: Vibration, acoustic emission, and ultrasonic sensors detect mechanical wear in bearings, pumps, and motors.
>
> Process Parameters: Temperature, pressure, flow, and pH sensors provide a continuous stream of quality-defining data.
>
> Environmental & Energy: Sensors track ambient temperature, humidity, and real-time energy consumption (kilowatt-hours) per machine or line.
>
> *Information Technology (IT) Data: The Record of Business.*
>
> Enterprise Resource Planning (ERP): Systems like SAP or Oracle hold transactional gold: master data (materials, suppliers, customers), purchase orders, production orders, inventory levels, and financial records.

Manufacturing Execution Systems (MES): This layer bridges ERP and the shop floor, providing detailed work orders, genealogy (tracking components through assembly), labor tracking, and as-built records.

Product Lifecycle Management (PLM) & Quality Management Systems (QMS): PLM systems hold CAD files, bills of materials (BOMs), and engineering change orders. QMS software manages non-conformance reports (NCRs), corrective and preventive actions (CAPAs), and supplier quality data.

Human-Generated & External Data: The Contextual Layer.

Human Input: Manual inspection reports, operator shift logs, maintenance work orders, and safety incident reports.

Supply Chain Data: Advanced Shipping Notices (ASNs) from suppliers, real-time freight tracking (GPS, container status), and supplier performance scorecards.

External Market Intelligence: This is a frontier for competitive advantage. Integrating anonymized point-of-sale (POS) data, social media sentiment analysis, commodity pricing feeds, and even weather forecasts allow AI models to correlate external factors with internal performance and demand.

2.1.2 Data Governance & Hygiene: From Chaos to Trusted Currency

Collecting data is only the beginning. Without rigorous governance, data rapidly decays into a costly liability. Data governance is the framework of policies, standards, and processes that ensure data is *Accurate, Consistent, Accessible, and Secure* throughout its lifecycle (Otto, 2011).

Establishing a Single Source of Truth: The most pernicious problem in manufacturing is conflicting data. Is the inventory count in the ERP, the WMS (Warehouse Management System), or the physical tally, correct? Governance mandates clear ownership and synchronization rules. For example, the MES is the "golden record" for production yield, and all other systems must be aligned to it.

Metadata and Data Lineage: Every data element must be tagged with metadata: What is it? Where did it come from? When was it captured? Who owns it? What are its units of measure? *Data lineage* tracks how data moves and transforms from its source to its consumption in an AI model, which is critical for debugging and regulatory compliance (e.g., in pharmaceuticals or food production).

Data Quality Rules and Automated Monitoring: Implement automated checks at the point of ingestion. Rules can flag:

Completeness: Is a required sensor reading null?

Validity: Is a temperature reading within a physically possible range (e.g., -50°C to 200°C for a process)?

Consistency: Does the "line stopped" event from the PLC align with zero production in the MES for the same timestamp?

Timeliness: Is the data arriving within its expected latency window?

Master Data Management (MDM): This is particularly crucial for mass customization. A centralized MDM system ensures that a "SKU #A123" has one established definition—its correct BOM, packaging specs, and quality parameters—across the ERP, MES, and PLM, preventing catastrophic errors on the production line.

2.1.3 Data Architecture: The Data Lake vs. Data Warehouse - A Practical Synergy

Where does all this consolidated data live? The choice of architecture determines the agility and cost of your AI initiatives.

The Data Warehouse: The Structured Repository for Business Intelligence. Optimized for storing and rapidly querying highly structured, cleansed historical data. It uses a predefined schema (like a relational database with strict tables for orders, customers, etc.). This is best for standardized operational reporting, dashboards for executives, and analyzing known business questions (e.g., "What was our OEE by line last quarter?").

Data Warehouse have limitations for AI. It is inflexible. Ingesting new, unstructured data streams (like machine vibration logs or images) is slow and expensive. It is not designed for the exploratory, "sandbox" nature of AI development.

The Data Lake: The Raw Reservoir for Discovery and AI. It is a vast, scalable repository (often on cloud platforms like AWS S3 or Azure Data Lake) that stores raw data in its native format—structured, semi-structured (JSON, XML), and unstructured (images, audio, text logs). It is best for foundational layer for AI/ML. Data scientists can "land" all available data here without upfront transformation, explore it freely, and prepare specific datasets for model training. It supports the agility needed for innovation (Fang, 2015). Its data set requires the same rigorous metadata tagging, cataloging, and access controls as a warehouse to be usable.

The Modern Hybrid Architecture: The most effective approach for AI-driven manufacturing is a layered architecture:

> Data Lake (Raw): All source data is ingested and stored here.
>
> Data Lake (Curated): A governed zone where raw data is cleaned, transformed, and enriched according to business rules, creating trusted datasets.
>
> Data Warehouse / Data Marts: Specific subsets of the curated data are loaded into warehouses or smaller "marts" for high-performance business intelligence and reporting.
>
> ML Feature Store: A specialized repository that stores pre-computed, reusable "features" (data attributes) optimized for ML models, accelerating data science work.

2.2 The Nervous System: IoT and Industrial Connectivity

If data is the fuel, then the network is the high-pressure fuel line delivering it to the AI engine. In a smart factory, connectivity is the central nervous system, enabling real-time communication between machines, sensors, gateways, and cloud platforms.

2.2.1 Retrofitting vs. Buying New: The Brownfield Conundrum

Most consumer goods manufacturers operate "brownfield" sites with a mix of legacy equipment (some decades old) and newer machinery. A pragmatic strategy is required.

Retrofitting Legacy Equipment: This is often the most cost-effective path.

Hardware Add-ons: Install "bolt-on" sensor kits (vibration, temperature) and industrial IoT (IIoT) gateways. These gateways (from companies like Siemens, Advantech, or Cisco) connect to machine PLCs via standard ports (Ethernet, serial) or even add-on cards, reading data and converting proprietary protocols to a standard like MQTT or OPC-UA.

Edge Computing Devices: Deploy ruggedized edge computers directly on the factory floor. They can pre-process data (e.g., performing Fast Fourier Transforms on vibration data to extract features), filter out noise, and only send meaningful insights to the cloud, reducing bandwidth costs and latency.

The Buy New Consideration: For mission-critical equipment or new production lines, specify *connected by design*. Procurement criteria should mandate native OPC-UA servers, embedded sensor arrays, and cybersecurity certifications (like IEC 62443) as non-negotiable features, not optional extras.

2.2.2 Network Architecture: Designing for Scale and Speed

The factory network can no longer be a single, flat IT network. It requires segmentation and technology tailored to different needs.

The Purdue Model/ISA-95 as a Blueprint: This industry-standard model provides a conceptual framework for network segmentation, separating layers from Level 0 (sensors) to Level 5 (enterprise IT). The key is a *Demilitarized Zone (DMZ)* between the Operational Technology (OT) (Levels 0-3) and IT (Levels 4-5) networks to control data flow and provide a security buffer.

Wireless Technologies for Flexibility:

Wi-Fi 6/6E: Provides high-bandwidth, reliable connectivity for mobile devices (tablets, AR headsets), AGVs (Automated Guided Vehicles), and handheld scanners. Its improved efficiency supports dense deployments of IoT devices.

Private 5G: A game-changer for industrial applications. It offers ultra-reliable low-latency communication (URLLC), massive machine-type communication (mMTC), and seamless mobility. It is ideal for coordinating fleets of autonomous mobile robots (AMRs) or for mission-critical control where a dropped Wi-Fi signal is unacceptable (Li et al., 2021).

The Critical Role of Edge Computing: Sending all sensor data to a centralized cloud for analysis is often inefficient and slow. *Edge computing* places processing power physically close to the data source.

Use Case - Real-Time Vision: An edge device with a graphics processing unit (GPU) can run a computer vision model directly on a camera feed to inspect 1,000 bottles per minute, making a pass/fail decision in milliseconds and triggering a reject arm—all without sending a single image to the cloud.

Use Case - Latency-Sensitive Control: An edge controller can process sensor data from a high-speed packaging line and make micro-adjustments to maintain registration, where even 100ms of cloud latency would be too slow.

2.2.3 Cybersecurity: Securing the Expanding Attack Surface

Connecting Operational Technology (OT) networks to Information Technology (IT) systems and the internet exposes previously isolated industrial control systems to a universe of digital threats. A cyber-attack can now cause physical damage, production stoppages, and safety hazards (Stouffer et al., 2015).

The OT Security Mindset: OT security prioritizes *safety and availability* over confidentiality. Patching a critical Programmable Logi Controller (PLC) may require a planned production shutdown, unlike an IT server. Security strategies must be adapted to this reality.

Essential Security Frameworks & Practices:

Zero Trust Architecture: Assume no entity (user, device, application) inside or outside the network is trustworthy. Verify explicitly and enforce least-privilege access for every access request.

Network Segmentation: Strictly segregate networks using firewalls and VLANs. The packaging line network should not be able to talk directly to the R&D network.

Asset Inventory & Vulnerability Management: Maintain a real-time inventory of every connected device (OT and IT). Continuously scan for known vulnerabilities and have a risk-prioritized patching schedule.

Continuous OT Network Monitoring: Deploy specialized tools that understand industrial protocols (like Claroty, Dragos, or Nozomi Networks). They passively monitor network traffic to detect anomalies (e.g., a PLC suddenly initiating communication to an unknown IP address), which could indicate a malware infection or unauthorized access.

2.3 The Human Foundation: Cultivating an AI-Ready Culture

Technology and data are inert without people to wield them. The single greatest point of failure in digital transformation is cultural resistance. Success requires proactively shaping an organizational culture that embraces change, trusts technology, and focuses on human-machine collaboration (Brynjolfsson & McAfee, 2014).

2.3.1 Transparency and Trust: Disarming Fear with Communication

The specter of job loss is the most potent source of resistance. Leaders must address this head-on with honesty and a compelling vision for the future of work.

Framing AI as Augmentation, Not Replacement: Consistently communicate that AI's primary role is to eliminate dull, dirty, and dangerous tasks—the tedious data entry, the repetitive visual inspection, the hazardous material handling. The goal is to elevate the human role.

The New Value Proposition for Employees: Articulate the new, more valuable work AI enables:

> The Maintenance Technician becomes a *Reliability Engineer*, analyzing AI-driven predictions to plan sophisticated interventions.
>
> The Quality Inspector becomes a *Process Improvement Analyst*, using defect trend data from vision systems to solve root-cause problems.
>
> The Line Operator becomes a *System Orchestrator*, managing the flow of work between automated cells and handling complex exceptions.

"Glass-Box" vs. "Black-Box" AI: Whenever possible, choose or design AI systems with Explainable AI (XAI) features. An operator should be able to understand why a

model flagged a product as defective (e.g., "90% confidence due to detected scratch in region X of the image"). This builds trust by making the AI's reasoning interpretable (Gunning et al., 2019).

2.3.2 Change Management: A Phased, Inclusive Approach

A top-down, forced implementation will fail. Successful adoption requires involving the end-users as co-creators in the process (Kotter, 2012).

Phase 1: Co-Design and Pilot. Form cross-functional teams of IT, engineering, and *frontline operators* to select pilot projects. Operators provide the crucial domain expertise on workflow pain points. They help design the human-AI interface and define success metrics.

Phase 2: Supervised Adoption and Feedback. During pilot rollout, provide intensive support. Create "super-users" or "digital champions" on the floor—respected operators who learn the system first and become peer trainers. Establish formal, easy channels for feedback on tool usability and functionality.

Phase 3: Scale with Refined Processes. Use learnings from the pilot to refine the technology and the accompanying business processes before scaling to other lines or plants.

Celebrate and broadcast the pilot's successes, showcasing the tangible benefits to both the business and the employees involved.

2.3.3 The Leadership Mandate: From Sponsor to Evangelist

Digital transformation is not an IT project; it is a business-wide strategic initiative that requires unwavering commitment from the highest levels of leadership.

Articulating the "Why" and the "What's In It For Us": The CEO and plant leadership must be the chief storytellers, constantly linking AI initiatives to the company's core mission, competitive threats, and opportunities for growth. They must connect the dots for every employee.

Allocating Resources and Accepting Risk: Leadership must provide not just budget for technology, but also time and space for learning and experimentation. They must champion pilot projects, accepting that some will fail, and frame those failures as valuable learning, not waste.

Modeling the Behavior: Leaders must themselves engage with the new tools and data. If the plant manager relies on a daily printed report instead of the real-time AI-powered dashboard, the message to the organization is clear: this isn't important.

Conclusion: The Foundation is the Future

Building the digital foundation is arduous, unglamorous work. It involves wrestling with legacy systems, defining data policies, running cable, and having countless conversations to assuage fears and build buy-in. However, this chapter has argued that this work is not a prerequisite—it is the transformation itself.

A company that has mastered its data, built a secure and agile network, and fostered a culture of trust and continuous learning is already fundamentally different from its competitors. It possesses the organizational "muscle memory" to adapt, learn, and innovate. When the next wave of technology arrives, this company will not need another "transformation"; it will simply evolve. Therefore, investing in data, connectivity, and culture is not just building a foundation for AI; it is building the resilient, intelligent, and human-centric factory of the future.

AI IN ACTION – TRANSFORMING CORE OPERATIONS

CHAPTER 3: INTELLIGENT PRODUCTION & PROCESS OPTIMIZATION: ENGINEERING THE SELF-OPTIMIZING FACTORY

Introduction: From Automated to Autonomous Operations

The promise of Industry 4.0 culminates in the vision of the "lights-out" factory, but a more immediate and attainable goal is the *self-optimizing production system*. This chapter moves beyond foundational technologies to explore their applied integration at the very heart of manufacturing: the orchestration of maintenance, scheduling, and quality control. Here, Artificial Intelligence (AI) transitions from an analytical tool to an active participant in the control loop, creating processes that are not only intelligent but also inherently resilient and adaptive.

For the operator, this represents the most tangible shift in daily reality. The role evolves from one of constant manual intervention and reactive problem-solving to one of overseeing and guiding an intelligent system. This chapter details how AI transforms three core operational pillars—maintenance, planning, and quality—from cost centers and sources of variability into engines of reliability, efficiency, and guaranteed excellence. We will examine the technical architectures, the data flows, and the profound economic impacts of creating a production environment that learns from itself and proactively engineers its own performance.

3.1 Predictive & Prescriptive Maintenance: From Cost Center to Reliability Engineering

Traditional maintenance strategies exist on a spectrum from reactive (run-to-failure) to preventive (time-based schedules). Both are economically suboptimal: the former causes catastrophic downtime and secondary damage, while the latter leads to unnecessary parts replacement and "infant mortality" failures from improper reinstallation. AI-driven Predictive and Prescriptive Maintenance (PdM/PxM) represents a paradigm shift, transforming maintenance from a tactical expense into a strategic function focused on asset health and system reliability (Selcuk, 2017).

3.1.1 Moving from Reactive to Predictive: The Science of Failure Forecasting

The core of PdM is the application of Machine Learning (ML) to time-series sensor data to identify the unique "fingerprint" of incipient failure.

Data Acquisition & Feature Engineering: The process begins with instrumenting critical assets. Vibration analysis, thermography, ultrasonic testing, and motor current signature analysis provide rich, high-frequency data streams. For a centrifugal pump on a beverage filling line, key sensors might include:

1. Triaxial accelerometers to capture vibration spectra in three planes.
2. Infrared sensors to monitor bearing and casing temperature.
3. Pressure transducers at inlet and outlet.
4. Flow meters to correlate performance with condition.

Raw sensor data is transformed into features—calculated metrics that are indicative of health. For vibration, this includes overall vibration level, spectral bands (e.g., specific frequencies associated with bearing ball-pass defects), and time-domain features like kurtosis, which detects impulsive signals from early-stage pitting (Jardine, Lin, & Banjevic, 2006).

Model Training & Failure Signature Development: Supervised ML models, such as Random Forests or Recurrent Neural Networks (RNNs), are trained on historical data. This training dataset must include both normal operation data and, crucially, data leading up to known, documented failures. The model learns the subtle, multivariate patterns that precede different failure modes (e.g., imbalance vs. misalignment vs. bearing wear). It establishes a *health baseline* and learns to calculate a *Remaining Useful Life (RUL)* probability distribution, providing a forecast, such as "90% probability of failure within the next 120-150 operating hours" (Lei et al., 2018).

The Human-Machine Interface: From Alarms to Insights: Instead of a simple high-limit alarm, the system provides a dashboard with asset health scores, trended RUL estimates, and diagnostic insights (e.g., "Primary fault indicator: rising amplitude at 3.2x shaft frequency, consistent with inner race bearing defect"). This shifts the technician's role from emergency responder to planned intervener.

3.1.2 Prescriptive Actions: Closing the Loop with Intelligent Recommendations

Predictive analytics answers "what will fail and when?" Prescriptive analytics answers "what should we do about it?" It integrates maintenance planning with broader operational constraints to recommend optimal actions (Kobbacy & Murthy, 2008).

Multi-Constraint Optimization Engine: A PxM system doesn't operate in a vacuum. It connects to:

1. ERP/MES: To understand production schedules, planned downtime windows, and criticality of the asset to upcoming orders.

2. CMMS/EAM: To check spare parts inventory, technician skill sets, and availability.

3. Financial Systems: To weigh the cost of a potential failure (lost production, scrap, secondary damage) against the cost of intervention (parts, labor, planned downtime).

Generating the Prescription: The engine processes all constraints to generate a ranked set of actionable recommendations:

1. Optimal Intervention Window: "Schedule repair during the 4-hour planned sanitation break next Thursday."

2. Required Resources: "This repair requires a Level II mechanic and part #BP-2043, which is in stock at the central warehouse."

3. Detailed Work Instructions: "Based on the fault diagnosis, the work order should include steps for inspecting and replacing the rear bearing assembly, followed by laser shaft alignment to within 0.002 inches."

4. Deferred Action Analysis: "If we cannot repair until Friday, the predicted increase in failure probability is 15%, with an associated production risk cost of $X."

Integration with Logistics: In advanced systems, the prescription can automatically generate a pick list for the warehouse robot to kit the required parts and tools, delivering them to the job site just-in-time for the technician.

3.1.3 ROI Case Study: Transforming a Contract Packaging Operation

Background: A mid-sized contract packaging plant for home care products operated with a preventive maintenance schedule and experienced an average of 15% unplanned downtime annually on its high-speed liquid filling and capping lines, leading to frequent overnight premium labor charges and missed delivery windows.

Implementation: The company deployed a vendor-agnostic PdM/PxM platform. Vibration sensors and thermal cameras were installed on 12 critical assets (fillers, cappers, conveyors, and homogenizer motors). The platform ingested two years of historical work order data to train initial models.

Results after 18 Months:

Unplanned Downtime Reduction: Decreased by 40%, from 15% to 9% of total operating time. This translated to over 500 additional production hours annually.

Maintenance Cost Reduction: Overall maintenance costs fell by 25%. This was achieved through a 60% reduction in emergency parts expediting fees, a 30% reduction in overtime labor, and a 15% increase in mean time between repairs (MTBR) for targeted components.

Secondary Benefits: Product waste from line jams and sudden stops decreased by 8%. The maintenance team's productivity shifted, with 70% of work now being planned versus 30% previously.

Key Insight: The most significant financial return came not from predicting a single catastrophic failure, but from the aggregation of dozens of smaller, sub-critical interventions that prevented cascading micro-stoppages and quality deviations, smoothing overall production flow.

3.2 AI-Driven Production Scheduling & Dynamic Planning: Solving the Impossible Puzzle

In an era of mass customization and volatile demand, production scheduling is no longer a linear, weekly planning exercise. It is a continuous, high-stakes optimization problem with thousands of dynamic variables. AI schedulers act as digital production orchestration engines, capable of finding feasible, optimal solutions in real-time that are beyond human calculation (Mourtzis, 2020).

3.2.1 The Multi-Variable Puzzle: Simultaneous Constraint Optimization

A modern AI scheduler treats the factory not as a collection of independent lines, but as a single, interconnected system. It ingests and balances a vast array of constraints:

> *Demand Constraints:* Customer due dates, priority levels, and penalties for late delivery.
>
> *Capacity Constraints:* Machine capabilities, speeds, and tooling requirements for each SKU.
>
> *Sequence-Dependent Setup Times:* The time and material waste to change from producing "Lemon

Scent" detergent to "Ocean Breeze" is different than changing to "Unscented."

Material & Resource Constraints: Real-time inventory of raw materials, packaging components, and availability of skilled operators or technicians for specific tasks.

Energy & Utility Constraints: Optimizing schedules to shift high-energy processes to off-peak tariff periods.

Preventive Maintenance Windows: Integrating the prescribed maintenance schedules from the PxM system.

Using constraint programming and genetic algorithms, the AI scheduler evaluates millions of potential sequences in minutes, finding the schedule that maximizes a defined business objective, such as Overall Equipment Effectiveness (OEE), on-time delivery, or gross margin (Pinedo, 2016).

3.2.2 Scenario Planning ("What-If" Analysis) and Dynamic Replanning

Static schedules break upon contact with reality. The true power of AI scheduling lies in its dynamic response capability.

Proactive Simulation: Planners can run "what-if" scenarios:

> "If we accept this high-priority rush order for Brand X, what is the impact on our other commitments for this week?"

> "If the resin delivery is delayed by 48 hours, how should we re-sequence the following five days of production to minimize disruption?"

The system instantly simulates each scenario, providing a visual impact analysis on Gantt charts and quantifying trade-offs in terms of delayed orders, increased changeovers, or overtime costs.

Autonomous Re-optimization: When a disruption occurs—a machine breakdown, a quality hold on a raw material batch, an operator absence—the AI scheduler does not require a manual replanning meeting. It automatically:

> Recognizes the constraint violation (e.g., "Filler Line 2 unavailable for 4 hours").

> Reserves the necessary capacity for repair.

Re-optimizes the remaining schedule for all other lines and work centers, considering all current constraints.

Pushes updated work instructions to the MES and floor displays, and notifies affected stakeholders (e.g., "Order #4567 for Brand Y has been moved from Line 2 to Line 5, new start time 14:30").

3.2.3 Yield Optimization: Closed-Loop Process Control

Beyond sequencing, AI enables real-time optimization of the production process itself to maximize the output of saleable product—the First Pass Yield (FPY).

Process Parameter Correlation: In complex processes like chemical mixing, baking, or plastic extrusion, final product quality (viscosity, moisture content, tensile strength) is a function of numerous input parameters (ingredient ratios, temperatures, pressures, screw speeds). Traditional Statistical Process Control (SPC) monitors outputs but reacts slowly.

AI-Driven Adaptive Control: An ML model, trained on historical process and quality data, learns the complex, non-linear relationships between inputs and outputs. In real-time, it:

> Monitors all input parameters and intermediate product states.
>
> Predicts the final quality outcome based on the current trajectory.
>
> Prescribes micro-adjustments to keep the process within the optimal "golden zone." For example, in a snack food fryer, the AI might adjust heat zones and conveyor speed in response to real-time moisture sensor readings of the incoming dough to ensure perfect color and crispness, reducing off-spec product by 3-5% (Qin & Badgwell, 2003).

Digital Twin Integration: The most advanced implementations use a process digital twin—a high-fidelity, physics-based simulation of the production line. The AI controller tests proposed adjustments in the virtual twin before executing them in the physical world, enabling safe, rapid optimization of even novel product runs.

3.3 The Autonomous Quality Gateway: From Detection to Prevention

Quality control has historically been a bottleneck—a sampling-based inspection process that occurs *after* production, catching defects too late to prevent waste. AI transforms quality into a pervasive, proactive, and preventative system that is integrated into every step of the value stream.

3.3.1 Computer Vision for 100% Inspection: Superhuman Consistency

Automated Visual Inspection (AVI) powered by deep learning represents one of the most mature and impactful AI applications in manufacturing.

System Architecture: A typical AVI station consists of high-resolution, high-speed industrial cameras (often line-scan or area-scan) with controlled lighting (backlight, darkfield, coaxial) to highlight features of interest. Images are processed by an edge computing appliance running a convolutional neural network (CNN).

The Power of Deep Learning: Unlike rule-based vision systems that look for specific pixel patterns, a CNN is trained on thousands of labeled images of "good" products and all known defect types (scratches, dents, misprints, missing components, cap torque issues). It learns to generalize, often identifying novel defect patterns or subtle anomalies that were not in the training set but fall outside the learned distribution of "good." This enables defect detection rates exceeding 99.9%, far surpassing human visual acuity and consistency, which deteriorates due to fatigue (Deng et al., 2019).

Beyond Pass/Fail: Defect Classification and Root-Cause Analytics: Advanced systems don't just reject a product; they classify the defect (e.g., "scratch-type B," "label misalignment >2mm"). This classified data, timestamped and linked to the machine and shift, flows into a quality data lake. Analytics dashboards then correlate defect spikes with specific machines, operators, or material batches, enabling rapid root-cause identification and correction.

3.3.2 Predictive Quality Control: Stopping Defects Before They Happen

Predictive quality moves the inspection upstream, using process data to forecast the quality of the final product while it is still being made.

In-Line Spectroscopy and Sensor Fusion: Technologies like Near-Infrared (NIR) spectroscopy can be deployed in-line to measure chemical composition, moisture, or coating thickness in real-time. An ML model correlates these real-time measurements with the final lab test results from that batch.

The Predictive Workflow:

During production of a cosmetic cream, in-line sensors monitor mixer torque, temperature, and homogenizer pressure.

The predictive quality model analyzes this data stream and calculates a Quality Confidence Score.

If the score drops below a threshold (indicating a high probability of final viscosity being out of spec), the system can either:

i. Alert the operator for immediate manual intervention.

ii. Automatically prescribe a corrective action (e.g., add 0.5% more emulsifier, mix for an additional 60 seconds).

iii. Automatically divert the suspect portion of the batch to a holding tank for rework, preventing it from contaminating the entire lot (Kourti, 2005).

Impact: This shifts quality from a *detective* to a *preventative* control, drastically reducing the cost of rework and scrap by catching deviations at their source.

3.3.3 Reducing the Cost of Non-Conformance: The Financial Imperative

The "Cost of Quality" is a well-established metric, with the "Cost of Non-Conformance" (CONC) being the most punitive component. It includes internal failure costs (scrap, rework) and external failure costs (returns, recalls, warranty claims, brand damage). AI-driven quality systems directly attack this cost (Juran & Godfrey, 1999).

Quantifiable Reductions:

> Scrap & Rework: 100% Automated Visual Inspection (AVI) and predictive quality can reduce scrap rates by 30-70% by catching defects immediately and preventing bad batches from being completed.

> Customer Returns & Chargebacks: For contract manufacturers, a major brand partner's chargeback for a quality escape can be devastating. AI inspection provides an auditable, digital quality record for every single unit shipped, virtually

eliminating escapes and the associated financial penalties.

Recall Prevention: The ability to trace a defect to a specific minute of production (via AVI and MES data) allows for surgical, targeted recalls of affected lots, rather than costly blanket recalls. Predictive quality makes recalls even less likely.

Brand Equity and Contract Security: Beyond direct costs, delivering demonstrably superior and consistent quality becomes a powerful competitive differentiator. It transforms the contract manufacturer from a commodity supplier to a strategic, zero-risk partner, securing longer-term contracts and premium pricing.

Conclusion: The Emergence of the Cyber-Physical Production System

Chapter 3 (of Part II) has illustrated how AI synthesizes data, connectivity, and intelligent algorithms to create a new class of production system: one that is self-aware, self-predicting, and self-optimizing. Predictive maintenance ensures the system's physical availability. AI-driven scheduling ensures its optimal utilization. The autonomous quality gateway ensures its output excellence.

Together, these intelligent processes create a virtuous cycle of improvement: Quality data improves process models, which improve scheduling accuracy, which reduces machine stress and improves maintenance forecasts. For the operator, the factory floor becomes less a place of manual control and more a theater of orchestrated intelligence, where human expertise is amplified by machine precision to achieve levels of performance, flexibility, and quality previously thought impossible. The foundation is now set for this intelligent production system to connect to an even larger cognitive network: the supply chain, which we will explore in the next chapter.

CHAPTER 4: THE COGNITIVE SUPPLY CHAIN: ORCHESTRATING RESILIENCE AND RESPONSIVENESS

Introduction: From Linear Pipeline to Neural Network

For decades, the supply chain has been conceptualized as a linear sequence of discrete, siloed functions: plan, source, make, deliver. This rigid model, often depicted as a static flow chart, is catastrophically brittle in a world defined by volatility, uncertainty, complexity, and ambiguity (VUCA). A single disruption—a port closure, a supplier bankruptcy, or a sudden demand surge—ripples through the chain, causing stockouts, excess inventory, and costly expedited freight.

Artificial Intelligence (AI) is fundamentally rewriting this construct. It enables the evolution from a linear supply chain to a *Cognitive Supply Chain (CSC):* a dynamic, self-learning, and adaptive network. In a CSC, information doesn't just flow; it is sensed, synthesized, and acted upon autonomously across nodes. The supply chain gains a form of situational awareness and collective intelligence, allowing it to anticipate disruptions, reconfigure itself in real-time, and optimize not just for cost, but for a balanced scorecard of service, resilience, and sustainability (Ivanov, Dolgui, & Sokolov, 2022).

This chapter details the transformation across three critical domains: how the chain senses demand, manages its sources, and executes physical movement. For the contract manufacturer, mastering these cognitive capabilities is no longer a competitive advantage—it is the price of admission to partnerships with leading, digitally-native brands.

4.1 Hyper-Accurate Demand Sensing & Forecasting: From Lagging Indicators to Leading Signals

Traditional demand forecasting is a rear-view mirror exercise, relying heavily on historical shipment or sales data. This approach fails in the face of new product launches, viral trends, or sudden shifts in consumer behavior. AI-powered demand sensing moves from extrapolation to real-time inference, creating a high-fidelity, short-term demand signal that is crucial for responsive manufacturing (Choi, Wallace, & Wang, 2021).

4.1.1 Moving Beyond Historical Sales: Integrating Multivariate Data Streams

The cognitive forecasting engine ingests and weights dozens of disparate, real-time data streams to detect demand signals often weeks before they appear in traditional sales reports.

Structured External Data:

> Point-of-Sale (POS) & E-commerce Data: Direct feeds from retail partners or online platforms provide near-real-time consumption data, not just warehouse withdrawals. For a contract manufacturer of snacks, seeing a 300% spike in POS sales for a particular chip flavor in a specific region on a Friday afternoon triggers a production adjustment for the following Monday.
>
> Syndicated Market Data: Services like Nielsen or IRI provide aggregated panel data on market share, pricing elasticity, and competitor promotions.
>
> Geospatial & Mobility Data: Foot traffic patterns near retail locations, derived from anonymized smartphone data, can predict demand uplift from local events, weather, or new store openings.

Economic & Commodity Indicators: Macroeconomic indices, disposable income trends, and raw material futures prices can signal broader demand trends for categories like consumer durables or premium goods.

Unstructured External Data (via NLP):

Social Media Sentiment & Trend Analysis: Natural Language Processing (NLP) models scan platforms like TikTok, Instagram, and Twitter. They don't just count mentions; they analyze sentiment, identify emerging influencers, and detect the "viral lift" for specific products, ingredients, or styles (e.g., "sea moss gel," "cottagecore aesthetic"). A spike in positive sentiment around a skincare ingredient can alert a cosmetics contract manufacturer to prepare for brand partner requests (Liu, 2012).

Search Engine & E-commerce Query Data: Trends in Google Search volumes or Amazon "clickstream" data for specific product terms are powerful leading indicators of intent to purchase.

News & Event Monitoring: NLP can scan news feeds for events that drive demand: an early heatwave (for beverages and sunscreen), a major sports event (for snacks and televisions), or a new health study (for related food supplements).

Internal Causal Data:

Promotional Calendars: Planned marketing campaigns, discounts, and advertising spend from brand partners.

New Product Introduction Plans: The launch trajectory and marketing support for new SKUs.

The AI model performs feature engineering on these streams, determining which combinations and lags are most predictive for different product categories, creating a constantly evolving, multi-dimensional demand picture.

4.1.2 Benefits for Contract Manufacturers: Operationalizing the Signal

For the contract manufacturer, a shared, hyper-accurate demand signal transforms planning from a guessing game into a collaborative, data-driven dialogue.

Confident, Agile Raw Material Procurement: Instead of placing bulk orders based on a quarterly forecast, procurement can move to a dynamic replenishment model.

The AI system generates a rolling, weekly requirement signal for raw and packaging materials. This allows for:

> Reduced Safety Stock: Higher forecast accuracy reduces the need for costly inventory buffers.
>
> Negotiation Leverage: Providing suppliers with a more reliable near-term forecast can improve pricing and allocation terms.
>
> Responsive "Just-in-Sequence" Procurement: Materials can be scheduled to arrive in alignment with the production sequence, minimizing warehouse dwell time.

Optimized Labor and Capacity Planning: Production schedulers gain visibility into expected demand volatility over the next 4-12 weeks. This enables:

> Pre-emptive Labor Planning: Scheduling temporary staff or planning overtime in advance of anticipated peaks.
>
> Strategic Capacity Allocation: Making informed decisions about which production lines to dedicate to which brand partners, and identifying when to initiate conversations about adding a third shift or investing in new equipment.

Efficient Changeover Scheduling: Grouping production runs for similar products to minimize cleaning and setup waste, based on a precise understanding of what needs to be made next.

Persuasive, Partnership-Oriented Client Discussions: Armed with AI-driven insights, the contract manufacturer shifts its role from a passive order-taker to a proactive strategic advisor. In capacity planning meetings with a brand, they can say: "Our models are detecting a significant uptick in online conversation and search volume for your new eco-friendly detergent pod. The data suggests a potential 40% upside to your initial forecast for Q3. We recommend we proactively reserve an additional 200 hours on Line 4 in August and secure a 15% larger allocation of the key bio-surfactant from our shared supplier." This builds immense trust and cements the manufacturer's position as an indispensable partner.

4.2 Smart Procurement & Supplier Management: From Transactional to Predictive

Procurement in the cognitive supply chain moves beyond cost negotiation and order placement. It becomes a function of risk intelligence and autonomous value capture, ensuring not just the lowest price, but the most resilient and economically optimal supply base.

4.2.1 Risk-Aware Sourcing: The Digital Supply Chain Immune System

Supplier risk is multidimensional. AI creates a continuous monitoring system that acts as an early-warning "immune system" for the supply network (Ivanov, 2022).

Financial Risk Monitoring: AI models scrape and analyze data from financial news, credit rating agencies, and regulatory filings to detect early signs of supplier distress—declining liquidity, mounting debt, or management turmoil—long before a bankruptcy filing.

Logistical & Operational Risk Monitoring: By integrating with global logistics data platforms (e.g., ship AIS tracking, port congestion reports, flight schedules), AI can model the health of supply routes. It can predict delays from events like a typhoon near a key Asian port or labor strikes at a European freight hub, prompting the system to qualify alternative shipping modes or routes.

Geopolitical & Compliance Risk Monitoring: Natural Language Processing (NLP) models monitor global news and government publications for geopolitical tensions, new trade tariffs, sanctions, or changes in environmental/safety regulations that could impact a supplier's ability to operate or export.

Performance Risk Analytics: Beyond delivery and quality scorecards, AI can analyze subtler patterns—increasing lead time variability, rising rates of minor non-conformances, or changes in communication responsiveness—that may indicate underlying operational problems at a supplier.

The Risk Dashboard & Prescriptive Actions: All this intelligence is synthesized into a Supplier Risk Index for each vendor. The system doesn't just flag a high-risk supplier; it prescribes mitigation actions: "Supplier A in Region X now has a 'High Risk' rating due to port congestion and political unrest. Recommended actions: 1) Trigger a pre-approved order with alternate Supplier B for 30% of needed volume. 2) Expedite current in-transit orders via air freight. 3) Schedule a risk-review meeting with the Supplier A account manager."

4.2.2 Dynamic Procurement: Autonomous Execution for Market Advantage

For commodity or semi-commodity materials (e.g., resins, oils, sugars, metals), AI enables a shift from periodic tendering to continuous, opportunistic buying.

Algorithmic Trading for Physical Goods: Similar to financial trading algorithms, AI procurement bots can be given a set of business rules: "Maintain a 4-week inventory of HDPE resin. When the spot price falls below $X per metric ton AND the supplier's quality rating is >Y, execute a purchase for up to Z tons." The bot monitors multiple market data feeds and e-commerce exchanges 24/7, executing transactions instantly when conditions are met, capturing cost savings no human team could reliably achieve.

Multi-Source Optimization: For critical materials, the AI system dynamically allocates orders across a pre-qualified supplier network. It continuously evaluates a total landed cost model that includes not just unit price, but also freight cost, tariff implications, and payment terms, automatically routing each purchase order to the optimal supplier at that moment.

Contract Compliance & Anomaly Detection: AI monitors all executed purchases against contract terms, instantly flagging invoicing errors, price discrepancies, or off-spec delivery quantities. This closes the loop on procurement leakage and ensures full value capture from negotiated agreements.

4.3 Autonomous Logistics & Warehouse Management: The Physical Flow of Intelligence

The cognitive supply chain's intelligence must manifest in the physical world. In the warehouse and on the road, AI and robotics create a seamless, efficient, and self-optimizing flow of goods from production line to customer doorstep.

4.3.1 Intelligent Warehouse Bots: The Swarm in the DC

The traditional picker-to-parts warehouse is inefficient. Autonomous Mobile Robots (AMRs) transform it into a dynamic, parts-to-picker system.

Fleet Intelligence & Real-Time Optimization: A central Warehouse Execution System (WES) powered by AI does not pre-program fixed paths for robots. Instead, it manages the entire fleet as a coordinated swarm. Each robot communicates its location, battery level, and current task. The AI, aware of all pending orders, constantly re-computes the global optimal plan (Azadeh, De Koster, & Roy, 2019):

> Dynamic Task Allocation: Which robot is closest to the needed item and has the right attachment (e.g., a lift, a conveyor top)?

Traffic Flow Optimization: The system calculates paths to avoid congestion, much like a real-time traffic navigation app for a city of robots. It can even impose one-way flows in aisles during peak picking times.

Pick Path Sequencing: For multi-item orders, it sequences the retrieval of items in the most efficient geographical order, minimizing travel time.

Human-Robot Collaboration (Cobotics): AMRs bring shelves (or totes) to stationary human pickers at ergonomic workstations. The AI directs the sequence of pods to maximize the picker's productivity, while a screen at the station displays exact pick instructions and uses computer vision to verify the correct item was selected. This can increase pick rates by 200-300% while reducing physical strain and error.

Adaptive Storage: The AI can analyze item velocity data (how frequently an SKU is picked) and dynamically re-slot the warehouse. Fast-moving items are automatically repositioned to locations that minimize robot travel time, a process that continuously optimizes itself.

4.3.2 Predictive Shipping & Last-Mile Optimization: The Calculus of Delivery

The final leg of the supply chain—shipping—is often the most expensive and variable. AI brings predictive precision and multi-objective optimization to logistics.

Intelligent Carrier & Route Selection: For each outbound shipment, the AI evaluates a massive decision matrix in seconds:

> Carrier Performance Data: Historical on-time delivery rates, damage rates, and service quality for each carrier on each lane (e.g., "Carrier X is 95% on-time from Chicago to Dallas but only 70% from Chicago to Miami").
>
> Real-Time Network Data: Current carrier capacity, known delays (weather, hub congestion), and spot market rates.
>
> Multi-Modal Options: The cost and time trade-offs between ground, air, LTL (Less-Than-Truckload), FTL (Full-Truckload), and even emerging options like crowd-sourced delivery.
>
> Sustainability Metrics: The calculated carbon footprint for each routing option.

The system selects the optimal carrier and service

level based on a weighted score of cost, speed, reliability, and carbon impact, specific to the customer's requirements (e.g., "prioritize cost" vs. "must arrive in 2 days").

Predictive Delivery Management & Customer Communication: By ingesting real-time GPS telematics from carriers, along with traffic and weather data, the AI can predict the Estimated Time of Arrival (ETA) with high accuracy, dynamically updating it throughout the journey. This allows for:

Proactive Exception Management: If a delay is predicted, the system can automatically notify the customer and the recipient's receiving dock.

Dynamic Last-Mile Routing: For delivery fleets, the AI can optimize the final sequence of stops in real-time based on traffic, customer availability windows (e.g., "deliver between 4-6 PM"), and new pickup requests.

Automated Proof of Delivery (POD) & Invoicing: Computer vision on a driver's mobile device can capture a delivery signature or a photo of the delivered goods, automatically closing the shipment in the system and triggering the invoice.

Conclusion: The Networked Enterprise

Chapter 4 illustrates that the cognitive supply chain is not merely a set of better tools for planning, buying, and moving. It is a new organizing principle for the extended enterprise. The walls between the contract manufacturer's ERP, the brand partner's demand plans, the supplier's production schedules, and the carrier's tracking systems dissolve into a continuous flow of shared intelligence.

The contract manufacturer who successfully builds and participates in this cognitive network achieves a fundamental strategic advantage: unmatched operational resilience. They can absorb shocks, capitalize on opportunities, and deliver value with a speed and precision that siloed, linear competitors cannot match. They become not just a maker of products, but the intelligent, responsive, and indispensable nexus of their customers' supply ecosystems. This sets the stage for the final, human-centric challenge: leading the organizational transformation required to steward this intelligent network, which we will explore in the next chapter.

CHAPTER 5: THE AUGMENTED WORKFORCE: REDEFINING HUMAN POTENTIAL IN THE COGNITIVE FACTORY

Introduction: The Human-Machine Symbiosis

The narrative of technological progress has long been shadowed by the specter of human obsolescence. The advent of Artificial Intelligence (AI) in manufacturing has rekindled this anxiety, with visions of fully autonomous, "lights-out" factories seemingly leaving little room for the human worker. This chapter fundamentally reframes that narrative. AI is not a replacement for human intelligence; it is its most powerful amplifier. The true competitive frontier in the age of cognitive manufacturing lies not in the pursuit of full automation, but in the strategic and empathetic design of human-AI symbiosis.

The goal is to create an Augmented Workforce, where AI handles the tasks of computation, pattern recognition, and prediction at superhuman scale and speed, freeing human workers to excel at what they do best: critical thinking, contextual judgment, creative problem-solving, and social collaboration (Daugherty & Wilson, 2018). This chapter provides a blueprint for this transformation, moving beyond the technology itself to focus on its most critical component: the people who will wield it. We will explore how AI acts as a ubiquitous assistant, how organizations must strategically shepherd their talent through this transition, and how performance management must evolve to measure and cultivate the uniquely human skills that now hold the greatest value.

5.1 AI as the Ultimate Assistant: Redesigning the Workday

For the frontline operator, technician, and manager, AI integration should feel less like a disruptive takeover and more like the arrival of an indefatigable, omniscient colleague. Its role is to eradicate friction, reduce cognitive load, and amplify human expertise.

5.1.1 Assisted Decision-Making: From Guesswork to Guided Action

Traditional human decision-making on the factory floor is often based on experience, heuristics, and limited real-time data. This can lead to variability, slow response times, and suboptimal outcomes. AI-powered decision support transforms this into a precise, evidence-based process.

Context-Aware Intelligent Dashboards: Moving beyond static gauges, next-generation Human-Machine Interfaces (HMIs) are dynamic and prescriptive. For a process operator managing a bioreactor, the dashboard doesn't just show temperature (78°C) and pressure (1.2 bar). An integrated AI model, analyzing real-time viscosity readings, oxygen uptake rates, and historical batch data, provides a contextual recommendation: "Batch #A47-22 trending towards low yield. Recommended action: Increase thermostat to 80°C for 10 minutes to optimize enzyme activity. Predicted outcome: +5% final yield. Click to execute." The operator retains ultimate authority but is empowered with a data-driven insight that would be impossible to derive manually.

Prescriptive Analytics for Complex Troubleshooting: When a packaging line jams, the root cause can be one of dozens of interacting variables (film tension, seal bar temperature, product orientation). An AI troubleshooter, trained on thousands of past jam events, instantly analyzes sensor data from the preceding 30 seconds. It doesn't just alert to the jam; it prescribes a ranked list of probable causes and corrective actions for the technician: "Primary Likelihood (85%): Film feed motor encoder fault. Secondary (12%): Misaligned guide rail. Action: Inspect motor encoder wiring at Panel B, Connector J4." This turns minutes or hours of diagnostic work into a guided, focused intervention.

Predictive Workload Management for Supervisors: For shift supervisors, AI can predict staffing bottlenecks by correlating the production schedule with real-time absenteeism, break schedules, and individual certification data. It might recommend: "Forecasted bottleneck at the labeling station at 14:30. Suggested action: Reassign Operator *John Doe* (certified) from secondary packaging 15 minutes prior."

5.1.2 Knowledge Management & Training: Institutionalizing Expertise

Critical tribal knowledge often resides in the heads of veteran employees and is lost when they retire. AI, combined with Augmented Reality (AR), can capture, codify, and deliver this expertise instantly to any worker.

AR-Guided Procedures & Maintenance: A technician wearing AR smart glasses approaches a malfunctioning rotary filler. Instead of consulting a bulky manual or a static PDF on a tablet, the glasses recognize the machine model via computer vision. They then overlay a digital work instruction directly onto the technician's field of view: animated, step-by-step holographic guides highlight which panel to remove, which bolt to turn, and the precise torque setting. The system can even recognize parts via AR and confirm the correct replacement component is being used, preventing errors.

Just-in-Time Micro-Training: When an operator is assigned to a new machine or process, AR can provide an interactive, guided walkthrough. As they look at different components, contextual "info pins" appear, explaining the function and safe operating parameters. This reduces formal training time from days to hours and provides persistent, on-the-job support.

Expert-in-the-Loop Remote Assistance: For a rare, complex fault, a frontline technician can initiate a remote assistance session. Using their AR glasses' camera, they can share their live point-of-view with a remote subject matter expert hundreds of miles away. The expert can then draw annotations, highlight components, and pull up schematics that appear directly in the technician's AR display, enabling collaborative problem-solving as if they were side-by-side (Porter & Heppelmann, 2017).

5.1.3 Voice & Chat Interfaces: The Conversational Factory

In hands-busy, eyes-busy environments, traditional interfaces (keyboards, mice, touchscreens) are impractical and unsafe. Natural Language Processing (NLP) enables a conversational layer over all operational systems.

Voice-Activated Control & Querying: An operator with greasy hands can simply state: "System, report overall equipment effectiveness for Line 3 since start of shift." The AI parses the query, fetches the data, and provides a spoken and visual summary. They can then command: "Log a minor vibration issue on the capper motor. Schedule for PM check tomorrow." This creates a seamless, frictionless flow of information and action.

AI-Powered Chatbots for Operational Support: A digital assistant, accessible via a tablet or wall-mounted screen, can answer a vast range of ad-hoc questions 24/7:

a. "What's the current inventory of 500ml amber bottles?"

b. "Show me the quality control results for the last 10 batches of SKU #X."

c. "What is the procedure for handling a chemical spill of Product Y?"

The chatbot retrieves information from the ERP, MES, QMS, and safety databases, providing instant answers and freeing up supervisors for more complex tasks.

Automated Logging and Reporting: Instead of manually filling out end-of-shift reports, a supervisor can narrate a summary: "Log shift summary. Line 2 achieved 92% OEE. One unplanned downtime event at 10:15 for a belt replacement, resolved in 22 minutes. Quality yield was 99.8%. No safety incidents." The NLP system structures this narrative into a formal digital report.

5.2 Strategic Workforce Transition & Reskilling: Investing in Human Capital

The augmentation of work inevitably changes the nature of jobs. A proactive, humane, and strategic approach to workforce transition is not just an ethical imperative but a critical business strategy to avoid a crippling skills gap and foster employee buy-in.

5.2.1 Mapping Future Roles: The Evolution of Manufacturing Jobs

Organizations must conduct a detailed analysis of how each role will be transformed, using a framework of Evolve, Create, and Reduce.

Evolving Roles (The Augmented Professional):

- *Maintenance Technician → Reliability Engineer:* Shifts from wrench-turning and scheduled part replacement to analyzing predictive maintenance dashboards, interpreting vibration spectra, and managing the health of a system of assets.

- *Quality Inspector → Process Improvement Analyst:* Moves from manual sampling and visual inspection to analyzing data from 100% automated vision systems, identifying statistical trends in defect data, and leading root-cause analysis projects to permanently eliminate quality issues.

- *Machine Operator → Production System Orchestrator:* Transitions from manually controlling a single machine to overseeing a cluster of collaborative robots and automated processes, handling exceptions, performing changeovers, and ensuring the seamless flow of materials and information.

Creating Roles (New Specializations):

- AI/ML Operations (MLOps) Specialist: Manages the lifecycle of AI models in production—monitoring for model drift, retraining with new data, and ensuring the seamless integration of model outputs with control systems.

- Digital Twin Engineer: Develops and maintains the high-fidelity virtual models of production lines used for simulation, optimization, and operator training.

- Human-Robot Collaboration (HRC) Designer: A hybrid role combining ergonomics, safety engineering, and UI/UX design to create safe, efficient, and intuitive interfaces between humans and robotic systems.

Managing Role Reduction: For highly repetitive, rules-based tasks (e.g., manual data entry, simple sorting), automation may reduce demand. Responsible transition involves attrition-based management (not replacing departing staff), internal mobility (prioritizing displaced workers for reskilling into evolving roles), and transparent communication about the long-term vision.

5.2.2 Building a Reskilling Pipeline: The Learning Organization

Investing in the current workforce is more sustainable and loyal than perpetually hiring for new skills. A multi-tiered learning ecosystem is required.

Foundational Digital Literacy: Mandatory training for all floor staff on data literacy (interpreting charts, understanding basic statistics), cyber-hygiene, and the core concepts of how AI and automation will support their work (World Economic Forum, 2023).

Technical Upskilling Pathways: Structured programs, often developed in partnership with local technical colleges or online academies (e.g., Coursera, Udacity), to teach in-demand skills:

> *For Technicians:* Basic Python for data analysis, fundamentals of IoT and sensor networks, introductory robotics programming.

> *For Engineers:* Advanced data science, machine learning fundamentals, cloud computing for industrial applications.

Experiential, On-the-Job Learning: The most powerful reskilling happens in context. Piloting new AI tools with a cohort of "digital champion" employees creates a group of internal experts. Job rotation programs and stretch assignments into new, technology-focused projects provide hands-on experience.

5.2.3 Inclusive Design: Technology for Everyone

If AI tools are designed only for the digitally-native, they will fail. Inclusive design ensures augmentation benefits the entire, diverse workforce.

Multi-Modal Interfaces: Offer information and control through voice, touch, gesture, and AR to accommodate different preferences and physical abilities.

Adaptive User Experience (UX): Systems should adapt to the user's demonstrated skill level. A novice might receive more detailed, step-by-step guidance, while an expert can access "power user" shortcuts and deeper data layers.

Cultivating a "Psychological Safety" Culture: Employees must feel safe to experiment with new tools, make mistakes, and ask "stupid" questions without fear of reprimand. This is foundational for adoption and learning (Edmondson, 2018).

5.3 Performance Management in an Augmented Environment: Measuring What Matters

When AI handles routine execution, the value of human work shifts from output to outcome and insight.

Performance management systems must evolve to measure and incentivize these new forms of value creation.

5.3.1 New Metrics for Success: Beyond the Hourly Rate

Legacy metrics like "units per hour" or "machine utilization" become insufficient and can even be counterproductive, encouraging behavior that conflicts with broader system goals (e.g., running equipment at suboptimal speeds to hit a target).

Quality-Centric Metrics:

First-Pass Yield (FPY): Percentage of product that meets specification the first time through the process, without rework or scrap. This incentivizes getting it right from the start.

Cost of Non-Conformance (CONC): Attributing the financial cost of waste, rework, and returns to specific processes or shifts, focusing teams on prevention.

Improvement & Initiative Metrics:

Improvement Suggestions Submitted/Implemented: Tracking and rewarding employees for identifying process bottlenecks or safety hazards.

AI Tool Adoption & Proficiency: Measuring effective use of new decision-support systems, not as surveillance, but to identify who needs more support and who can become a peer coach.

System Health & Resilience Metrics:

Mean Time To Restore (MTTR): How quickly a team can recover from an unexpected disruption, measuring problem-solving agility.

Proactive Intervention Rate: The percentage of maintenance work that is planned and predictive versus reactive, indicating a shift towards reliability engineering.

5.3.2 Continuous Feedback Loops: AI for Human Development

AI can analyze operational data not just to optimize machines, but to provide personalized, constructive feedback for employee development.

Personalized Performance Insights: An AI coach could analyze a reliability engineer's work and provide feedback:

"You successfully resolved the bearing issue on Compressor A 20% faster than the site average. Your diagnostic path aligned with the AI's top recommendation in 9 out of 10 steps. For even faster resolution, consider reviewing the vibration analysis module on common misalignment patterns."

Dynamic Training Recommendations: Based on an employee's role, performance data, and career aspirations, the system can recommend specific micro-courses or experiences: "To advance towards the Process Improvement Analyst role, we recommend completing the 'Root Cause Analysis with Statistical Tools' module this quarter."

Sentiment and Engagement Analysis (Used Ethically): With clear transparency and consent, NLP analysis of anonymized communication and feedback can help leaders gauge team morale, identify pockets of confusion about new tools, and proactively address concerns, fostering a more responsive and supportive management approach.

Conclusion: The Human-Centered Factory

The cognitive factory's ultimate success will not be measured by its degree of automation, but by its level of human empowerment and ingenuity. The most significant investment in the age of AI is not in silicon and software alone, but in the people who will design, manage, and evolve these intelligent systems. By framing AI as the ultimate assistant, strategically navigating workforce transitions with empathy and foresight, and redesigning performance systems to value human judgment and creativity, manufacturers can build organizations that are not only more efficient and resilient but also more engaging, equitable, and innovative. The future belongs not to the factory that replaces its people, but to the one that most effectively amplifies their potential.

IMPLEMENTATION, ETHICS, AND THE ROAD AHEAD

CHAPTER 6: THE AI IMPLEMENTATION JOURNEY: A STEP-BY-STEP FRAMEWORK

Introduction: From Aspiration to Operational Reality

The promise of Artificial Intelligence in manufacturing—unprecedented efficiency, predictive foresight, and cognitive augmentation—is compelling. Yet, for many organizations, the path from conceptual enthusiasm to scaled, operational reality remains fraught with uncertainty. Failed pilots, unexplainable "black box" models, and solutions that struggle to integrate with legacy infrastructure are common pitfalls that erode value and confidence (Davenport, 2018). This chapter provides a comprehensive, actionable framework for navigating the AI implementation journey, transforming it from a series of disconnected experiments into a disciplined, repeatable process for enterprise-wide value creation.

Moving beyond the hype, successful AI adoption is not primarily a technological challenge; it is a *strategic, operational, and organizational* one. It requires a methodical approach that aligns technical initiatives with concrete business outcomes, manages risk through iterative learning, and builds the institutional scaffolding necessary for sustainable scaling (Iansiti & Lakhani, 2020). This chapter delineates a three-phase framework—Assessment & Prioritization, Pilot Project Execution, and Scaling & Integration—each expanded with detailed methodologies, decision matrices, and real-world considerations. This framework is designed to guide manufacturing leaders from the initial diagnostic stage through to the establishment of a pervasive, governed AI capability that delivers continuous competitive advantage.

6.1 Phase 1: Assessment & Prioritization: Laying the Strategic Foundation

Before writing a single line of code or issuing a purchase order, the critical first phase involves a clear-eyed assessment of the organization's readiness, opportunities, and strategic posture. This phase is about applying rigor to the "where" and "why" of AI, ensuring that initial efforts are focused on domains of genuine business value and technical feasibility (Garbuio & Lin, 2019).

6.1.1 Process Mining & Opportunity Discovery: The Digital Diagnostic

The most effective AI initiatives are not technology in search of a problem, but precise solutions targeting well-understood operational constraints. Process mining serves as the foundational diagnostic tool for this discovery (van der Aalst, 2016).

Extended Methodology:

1. Data Harvesting & Model Creation: Utilize specialized process mining software (e.g., Celonis, UiPath Process Mining) to automatically ingest event log data from core systems like Manufacturing Execution Systems (MES), Enterprise Resource Planning (ERP), and warehouse management systems. This data, which records the "who, what, when" of every transaction and workflow, is used to construct an objective, as-is Digital Twin of Operations (DTO). This is not a physical twin, but a dynamic, data-driven model visualizing the real flow of orders, materials, and information.

2. Conformance Checking & Bottleneck Analysis: The DTO reveals the stark reality between documented Standard Operating Procedures (SOPs) and actual practice—call it a *super-advanced comprehensive value stream analysis plus*. Algorithms analyze the model to identify:

    Chronic Bottlenecks: Repetitive delays at specific process steps (e.g., quality approval queues averaging 4.2 hours).

Variant Analysis: The startling number of different paths taken to complete the same process (e.g., 47 unique paths for "order-to-ship," indicating extreme variability and tribal knowledge).

Compliance Gaps: Instances where mandatory steps (e.g., safety checks) are bypassed.

Resource Inefficiencies: Underutilization of high-value assets or persistent wait times for shared resources.

3. Quantifying the Prize: The key is to move from identifying bottlenecks to *quantifying their impact*. For example, process mining might reveal that Packaging Line B has a 40-minute average changeover time, 30% above standard, and that 12% of changeovers result in minor misalignment errors causing the first 15 minutes of production to be scrapped. The AI opportunity becomes clear: a computer vision system for automated changeover verification and a reinforcement learning model to optimize changeover sequence. *The potential ROI is now calculable: (Reduction in scrap + Increased line availability) = Direct financial value.*

6.1.2 Build vs. Buy vs. Partner Analysis: A Strategic Capacity Decision

With high-value targets identified, the next critical decision is determining the execution model. This is a strategic choice balancing speed, control, cost, and long-term capability building (Fountaine et al., 2019). *Table 1, Decision Framework,* provides basic yet essential guidance through a clear, methodical approach to making the critical choice of whether to buy, build, or pursue a partnership path.

Leaders are ultimately compensated for sound decisions, those that advance the organization's objectives. When stakeholders answer each key evaluation question objectively and factually and leaders make *unburdened decisions*—without vendor bias and without defaulting to the cheapest option—this framework is highly likely to guide them toward a decision that aligns with the company's long-term strategic goals and supports a successful execution model.

Expanded Decision Framework:

Model	Best For...	Pros	Cons	Key Evaluation Questions
BUY (SaaS/Off-the-Shelf)	Common, well-defined problems (predictive maintenance for standard motors, invoice processing). Need for rapid deployment and minimal internal data science headcount.	**Speed to Value:** Fastest implementation. **Lower Upfront Cost:** Operational expense model. **Proven & Supported:** Vendor maintains and updates the model. **Reduced Risk:** Known performance benchmarks.	**Limited Customization:** May not fit unique processes perfectly. **Vendor Lock-in:** Data and workflows can become tied to a platform. **Black Box:** Limited ability to modify or understand the underlying model. **Recurring Costs:** Can become expensive at scale.	Is this a generic or a proprietary/core process? Does the vendor's roadmap align with our needs? What are the total cost of ownership (TCO) and exit strategy?
BUILD (In-House)	Problems involving proprietary IP, highly unique processes, or where AI is a core future competency. Requires significant, secure data assets.	**Maximum Control & Customization:** Solution is tailored precisely to the need. **IP Ownership:** All algorithms and data remain in-house. **Internal Capability Development:** Builds crucial AI/ML skills within the team. **Long-Term Flexibility:** Easier to adapt and extend.	**High Initial Cost & Time:** Requires investment in talent, infrastructure, and time. **High Risk:** Potential for project failure or underperformance. **Talent Scarcity:** Difficult and expensive to hire and retain top AI talent. **Long-Term Maintenance Burden:** The team owns the full lifecycle.	Do we have (or can we acquire) the requisite data science and MLOps talent? Is this problem strategic enough to warrant building a dedicated capability? Do we have the executive patience for a longer development cycle?
PARTNER (with System Integrators/ Consultancies)	Complex problems requiring cross-system integration, change management, and deep domain expertise. Organizations seeking a middle path between buy and build.	**Accelerated Learning:** Leverage partner's experience and best practices. **Reduced Implementation Risk:** Partner shares project risk. **Integrated Solution:** Expertise in tying AI to existing ERP, MES, and OT systems. **Capability Transfer:** Can include upskilling components for internal teams.	**Cost:** Can be capital-intensive. **Dependency:** Risk of over-reliance on the partner. **Cultural Fit:** Must align with internal ways of working. **Knowledge Retention:** Ensuring expertise is transferred, not just rented.	Does the partner have proven domain expertise in *manufacturing AI*, not just generic AI? What is the explicit plan for knowledge transfer and eventual ownership? How does the partnership model align with our long-term strategic goals?

Table 1: Decision Framework

Recommendation: Most successful manufacturing AI programs adopt a hybrid portfolio approach. They *Buy* for common, non-differentiating capabilities (e.g., email automation), *Partner* for complex, high-impact integrations (e.g., plant-wide predictive quality), and selectively *Build* in areas that constitute a true source of long-term competitive advantage (e.g., proprietary material science optimization) (Fountaine et al., 2019).

6.2 Phase 2: Pilot Project Execution: Proving Value in a Controlled Environment

The pilot phase is the crucial "proof of concept" that translates strategy into tangible results. Its purpose is not to deliver enterprise-wide transformation, but to de-risk the broader initiative, generate evidence-based confidence, and create a playbook for future projects (Ries, 2011).

6.2.1 Selecting the Right Pilot: The Goldilocks Principle

The ideal pilot is "just right"—neither too trivial to be meaningless nor too complex to be unachievable.

Expanded Selection Criteria:

1. High, Measurable Impact: Target a bottleneck with a clear, quantifiable cost (e.g., a specific visual

inspection station where 3% of defects escape to customers, costing $500k annually in returns and warranty claims).

2. Contained Scope: Limit the project to a single production line, a specific part family, or one warehouse. This simplifies data collection, integration, and change management.

3. Data Availability: The process must already generate or be easily instrumented to generate the necessary structured and/or image data for model training.

4. Executive & Frontline Sponsorship: Identify a business leader who owns the pain point and a respected floor-level "digital champion" who can drive adoption and provide feedback.

5. Clear Success Metrics (OKRs): Define unambiguous, time-bound objectives (Doerr, 2018):

    Objective: Reduce escapee defects from Visual Inspection Station #5.

    Key Results: 1) Achieve 99.5% detection accuracy on known defect types within 6 months. 2) Reduce false positives by over 50% to maintain line speed. 3) Demonstrate a positive ROI with a payback period of <18 months.

6.2.2 The MVP Mindset: Iterative Learning and Agile Development

Adopting a Minimum Viable Product (MVP) mindset is paramount. The goal is to field a working solution quickly, learn from its real-world performance, and iteratively improve (Ries, 2011).

Expanded Execution Approach:

1. Data Acquisition & Pipeline Creation: Before model development, establish a robust, automated pipeline to collect, clean, and label data. For a visual inspection pilot, this involves capturing thousands of images of both good and defective parts, often requiring careful labeling by quality experts. Tools like cloud-based data labeling platforms (Scale AI, Labelbox) can accelerate this.

2. Model Development & "Cognifying" the Workflow: Develop the initial algorithm (e.g., a convolutional neural network for image classification). Crucially, design the human-in-the-loop workflow: How will the AI's prediction be presented to the operator? (e.g., a monitor highlighting a suspected defect). What action does the operator take? (e.g., confirm or override). How are overrides fed back into the system to retrain and

improve the model? This feedback loop is essential for continuous learning (Daugherty & Wilson, 2018).

3. Rigorous Testing & Validation: Move beyond standard accuracy metrics. Conduct shadow testing, where the AI model runs in parallel with the existing human process without affecting output, to benchmark its performance. Perform stress tests with edge cases and novel defect types. Validate that the model's performance is consistent across different lighting conditions, product variations, and operators.

4. Deploy, Monitor, Iterate: Deploy the MVP to the live line. Closely monitor not just the model's technical performance (accuracy, drift), but also human factors: Is the interface intuitive? Do operators trust the recommendations? Are workflows smoother? Use this feedback for rapid, agile iterations—improving the UI, retraining the model with new data, and optimizing the process. The pilot is successful only when it demonstrates a clear, measurable improvement in the defined business outcome and achieves user acceptance.

6.3 Phase 3: Scaling & Integration: Industrializing the AI Capability

Scaling is where many AI programs stall. Moving from a successful pilot to dozens of deployed models requires a shift from project-centric thinking to product- and platform-centric operations. This phase is about building the institutional "factory" for AI (Iansiti & Lakhani, 2020).

6.3.1 Technology Stack Integration: The Interoperability Imperative

An AI model in a sandbox delivers no lasting value. Its power is unlocked through seamless integration with the core operational technology (OT) and information technology (IT) stack.

Expanded Integration Architecture:

1. Data Layer Integration: Establish secure, bidirectional data flows. The AI model must consume real-time data from Programmable Logic Controllers (PLCs), sensors, and MES (e.g., machine parameters, product IDs). Its predictions or instructions must be fed back into these systems—triggering a work order in the Computerized Maintenance Management System (CMMS),

updating a production order in the ERP, or sending a set-point adjustment to a PLC.

2. API-First & Middleware Strategy: Mandate that all AI solutions expose robust, documented APIs. Use an Industrial IoT (IIoT) platform or integration middleware (e.g., MQTT brokers, Azure IoT Hub, AWS IoT Greengrass) as the central nervous system to manage connectivity, data normalization, and message routing between legacy OT equipment, new AI microservices, and enterprise IT systems. This avoids point-to-point integration spaghetti.

3. Edge vs. Cloud Deployment: Implement a hybrid architecture. Time-sensitive, low-latency inferences (e.g., real-time robotic collision avoidance) run on edge devices (industrial PCs, specialized gateways) on the factory floor. Model training, retraining, and large-scale analytics run in the cloud, where scalable compute and storage are available. A unified MLOps platform should manage models across this continuum (Sculley et al., 2015).

6.3.2 Creating an AI Center of Excellence (CoE): The Governance Engine

Sustainable scale requires central coordination to prevent chaos, ensure quality, and propagate learning. The AI

Center of Excellence is a cross-functional governing body, not a centralized development team (Davenport, 2018).

Expanded CoE Structure & Mandate:

1. Cross-Functional Composition: The CoE comprises representatives from IT/Data Engineering, Data Science, OT/Engineering, Business Unit Leaders, Process Excellence, Cybersecurity, and Legal/Compliance. This ensures all perspectives are considered.

2. Core Responsibilities:

    Governance & Standards: Establish and enforce enterprise standards for model development (e.g., coding practices, version control with Git), documentation (e.g., model cards detailing intended use, limitations, and fairness metrics), and deployment (e.g., via a centralized Model Registry) (Mitchell et al., 2019).

    Platform Management: Curate and manage the shared MLOps platform (e.g., MLflow, Kubeflow, Azure ML) that provides tools for experiment tracking, automated model training pipelines, and model deployment.

Knowledge Hub & Best Practices: Document and share learnings, reusable code components, and case studies across the organization. Run internal workshops and "lunch-and-learn" sessions.

Talent Development & Upskilling: Coordinate with HR on career paths for AI roles (e.g., ML Engineer, Data Analyst) and manage the rollout of the upskilling programs detailed in Chapter 5.

Ethical & Risk Oversight: Develop guidelines for responsible AI, including bias detection in training data, model explainability requirements for high-stakes decisions, and regular audits of model performance in production to detect drift or degradation (Raji et al., 2020).

Conclusion: The Journey to an AI-Operational Enterprise

The AI implementation journey is not a one-time project with a definitive end date; it is the cultivation of a new, core organizational competency—the ability to continuously convert data into decisive action and insight. By meticulously following the phased framework of strategic assessment, disciplined piloting, and industrialized scaling, manufacturers can systematically derisk their AI investments

(Iansiti & Lakhani, 2020). This journey transforms AI from a scattered set of tactical tools into a strategic, integrated capability that permeates operations, empowering the augmented workforce and driving a sustainable cycle of improvement, resilience, and innovation. The ultimate destination is an AI-operational enterprise, where intelligent systems and human expertise are seamlessly fused, enabling a level of operational excellence that defines the future of manufacturing competitiveness.

CHAPTER 7: NAVIGATING RISKS, ETHICS, AND RESPONSIBLE AI IN MANUFACTURING

Introduction: The Imperative for Ethical, Trustworthy, and Secure AI

The integration of Artificial Intelligence into the physical world of manufacturing elevates its consequences from the digital realm to the tangible, impacting worker safety, product quality, economic livelihoods, and environmental stewardship. Unlike purely commercial AI, a failure in an industrial context can result in catastrophic equipment damage, significant financial loss, or physical harm. This chapter moves beyond the technical and strategic blueprints of previous sections to confront the essential, non-negotiable challenges of governance, ethics, and risk that underpin sustainable and responsible AI deployment.

A cognitive factory's resilience and long-term license to operate are contingent not just on its efficiency, but on its fairness, transparency, security, and accountability. This chapter provides a comprehensive framework for navigating these critical dimensions. We will dissect the origins and mitigation of Algorithmic Bias, which can subtly encode discrimination into automated decisions; confront the "Black

Box" Problem, arguing that in high-stakes industrial environments, interpretability is not a luxury but a safety and compliance requirement. Finally, we will examine the converging frontiers of Data Privacy, Security, and Intellectual Property, where the very data that fuels AI becomes a vault of priceless and highly vulnerable assets. This sober analysis is not intended to stifle innovation but to fortify it, ensuring that the immense power of cognitive systems is wielded with the wisdom, foresight, and responsibility demanded of the industrial age.

7.1 Algorithmic Bias & Fairness: From Invisible Flaws to Systemic Consequences

Algorithmic bias occurs when an AI system produces systematically prejudiced outcomes due to erroneous assumptions or unfair patterns in its training data or design. In manufacturing, bias is often less about social demographics and more about operational history, but its impact on fairness, efficiency, and safety can be equally severe (O'Neil, 2016).

7.1.1 Understanding the Risk: The Anatomy of Industrial Bias

Bias is not merely a "bug" in the code; it is a reflection of historical realities, measurement limitations, and human oversight, encoded and amplified by scale.

Expanded Sources of Bias in Industrial AI:

1. Historical Data Bias (The "Garbage In, Gospel Out" Problem): An AI model trained on years of maintenance records may learn that "Machine A fails every 500 hours." However, if those records only contain work orders for catastrophic failures and ignore near-misses or minor adjustments made by experienced technicians, the model is biased toward late-stage, reactive maintenance. It perpetuates the suboptimal historical practice, missing the opportunity for earlier, predictive intervention.

2. Measurement & Labeling Bias: In visual inspection systems, the definition of a "defect" is subjective. If a training dataset is labeled primarily by one quality inspector with a particular tolerance threshold, the model will inherit that inspector's bias. It may consistently reject parts that another inspector—or a customer—would accept, leading to unnecessary scrap and cost. Similarly, sensor placement bias (e.g.,

only monitoring the center of a reactor vessel) creates an incomplete and biased view of the process.

3. Proxy Variable & Correlation Bias: An AI used for predictive hiring or promotion in a plant might be trained on historical data correlating "success" with attributes like "long tenure" or "consistent shift schedule." These can become proxies that unfairly disadvantage qualified candidates who have taken family leave, are new to the industry, or work flexible schedules, thereby perpetuating a non-diverse workforce and missing out on talent.

4. Automation Bias (Human-in-the-Loop): This is the human tendency to over-rely on automated recommendations. An operator, trusting a biased AI scheduler that always prioritizes "Line 1" for high-margin orders, may stop questioning the logic, leading to the chronic under-utilization and delayed maintenance of "Line 2," creating a self-fulfilling prophecy of its inferior performance.

7.1.2 Mitigation Strategies: A Proactive Framework for Fair AI

Combating bias requires a proactive, multi-disciplinary approach embedded throughout the AI lifecycle, from problem formulation to post-deployment monitoring (Mitchell et al., 2019).

Expanded Mitigation Framework:

1. Diverse & Inclusive Development Teams: The first line of defense is cognitive diversity. Teams comprising data scientists, process engineers, frontline operators, ethicists, and union representatives will ask different questions. An operator might immediately spot that a "recommended optimal speed" from an AI is unsafe under certain cleaning conditions, a nuance a remote data scientist could miss. This diversity challenges assumptions in problem framing and solution design.

2. Rigorous Data Auditing & Curation: Before training, data must be audited for representativeness and fairness.

   Demographic Parity Checks: For HR-related models, analyze whether outcomes (e.g., interview recommendations) are statistically independent of protected attributes (gender, ethnicity).

   Process Parity Checks: For operational models, analyze if data from all machines, shifts, and product variants are equally represented and of similar quality. Techniques like disparate impact analysis can quantify if a model's predictions disproportionately affect one asset class or production line.

3. Algorithmic Fairness Techniques & Bias-Aware Modeling: Integrate fairness constraints directly into the model development process.

   Pre-processing: Modify the training data to remove correlations between sensitive attributes (e.g., a specific machine ID) and the target variable.

   In-processing: Use algorithms that include fairness as an explicit optimization goal alongside accuracy (e.g., imposing a constraint that false alarm rates be equal across different product families).

   Post-processing: Adjust the outputs of a model after training to meet fairness criteria (e.g., calibrating confidence thresholds differently for different subgroups).

4. Continuous Monitoring for Bias Drift: Bias is not static. A model that is fair at deployment can become biased if the underlying process changes. Continuous monitoring for concept drift and prediction disparity across subgroups is essential. Establish a

schedule for regular fairness audits, retraining models with new, representative data.

7.2 Transparency, Explainability, and Trust: Illuminating the Black Box

In manufacturing, trust is earned through understanding. An operator will not follow an AI's shutdown command, a maintenance chief will not authorize a costly part replacement, and a regulator will not certify a process unless they can understand the *why* behind the AI's decision. Explainability bridges the gap between powerful prediction and actionable, trusted insight (Ribeiro et al., 2016).

7.2.1 The "Black Box" Problem: When Unexplainability Becomes Unacceptable

Deep neural networks and complex ensemble models are often inscrutable, making decisions based on millions of weighted connections. This opacity creates critical risks:

> *Safety & Compliance Risk:* If an AI-controlled robotic arm makes an unexpected movement, engineers must trace the cause to prevent recurrence. A black box offers no root cause.

Regulatory Risk: Industries like pharmaceuticals (governed by FDA 21 CFR Part 11) and aerospace require rigorous validation and documentation of any process affecting product quality or safety. "The model said so" is not a valid entry in a batch record or audit trail.

Adoption & Change Management Risk: Operators and engineers, trained to understand cause-and-effect, will reject or misuse a system they cannot comprehend, undermining its value and potentially creating unsafe workarounds.

Debugging & Improvement Risk: When a model's performance degrades, diagnosing whether the issue is bad data, a changing process, or a model flaw is nearly impossible without explanatory tools.

7.2.2 Explainable AI (XAI): Techniques for Building Interpretable Systems

The field of XAI provides a toolkit for making AI decisions more interpretable, ranging from using inherently interpretable models to applying post-hoc explanation techniques to complex ones.

Expanded XAI Methodology for Manufacturing:

1. The Interpretability Spectrum & Model Selection: The first choice is to use a simpler, inherently interpretable model when possible.

   High Interpretability: Linear models, decision trees, and rule-based systems. For example, a decision tree that prescribes maintenance can be read as a clear flow chart: "IF vibration > 5.2 mm/s AND temperature > 75°C, THEN flag for bearing inspection."

   Lower Interpretability (Requiring Post-hoc XAI): Deep learning, complex ensembles. Use these when their superior accuracy is critical (e.g., visual inspection of complex textures), and pair them with explanation techniques.

   2. Post-hoc Explanation Techniques:

   Local Interpretable Model-agnostic Explanations (LIME): For a single prediction (e.g., "this weld is defective"), LIME creates a simplified, interpretable model (like a linear regression) that approximates the complex model's behavior *just for that instance*. It might show: "This prediction was

85% based on the irregular pixel pattern in Region X of the image."

SHapley Additive exPlanations (SHAP): Based on cooperative game theory, SHAP assigns each input feature (e.g., pressure, temperature, feedstock lot) an importance value for a specific prediction. A SHAP summary plot can show that, globally, "batch viscosity" is the most important driver of final yield, but for a specific failed batch, "catalyst age" was the decisive negative factor.

Counterfactual Explanations: These provide actionable, "what-if" guidance. Instead of "weld rejected," the system could explain: "This weld would have been accepted if the travel speed had been reduced by 10%. Consider this adjustment for the next pass."

3. Designing for Explainability in the HMI: The technical explanation must be translated into an effective human-machine interface.

The "Five Whys" Dashboard: Layer explanations. The primary alert: "Predictive Maintenance: Pump P-101 - Impeller Fault Likely (92%)." First "Why?": "Top contributing factors: Vibration spike in axial plane, rising motor current." Second

"Why?": (Click) SHAP graph shows historical trend. Third "Why?": (Click) Counterfactual: "If vibration had remained below 4.0 mm/s, failure probability would be <10%."

Confidence Scores with Reasoning: Always present a confidence score alongside a recommendation, with a clear link to the evidence. "Recommended Set-point Change: Increase temperature to 155°C. Confidence: 78%. Basis: Correlates with successful past batches #A47, #A52 where initial pH was similar."

7.3 Data Privacy, Security, and Intellectual Property: Guarding the Cognitive Vault

In the cognitive factory, data is the new feedstock, and AI models are the proprietary recipes. This creates a profound shift in the asset portfolio, introducing unprecedented risks related to cyber-physical security, privacy law, and intellectual property (IP) theft (Anderson, 2020).

7.3.1 Protecting Proprietary Data and Models: A Defense-in-Depth Strategy

The attack surface expands from IT networks to operational technology (OT) and the AI pipeline itself.

Expanded Protection Framework:

1. Asset Identification & Classification: Catalog all AI-related assets: Training Datasets (the curated historical data), Operational Data Streams (real-time sensor feeds), Trained Model Files (the algorithms and weights), and Model Pipelines (the code for training and deployment). Classify them by sensitivity (e.g., "Crown Jewel" for a proprietary yield-optimization model).

2. Securing the AI/ML Pipeline (MLSecOps):

    Secure Development: Implement code signing, version control, and dependency scanning for model development code to prevent poisoning attacks.

    Data Provenance & Integrity: Use cryptographic hashing and blockchain-like ledgers to maintain an immutable record of data lineage—where training data came from, how it was transformed, and what model version it created. This is critical for auditability and detecting data tampering.

    Model Security: Protect model files at rest and in transit with strong encryption. Implement access controls so only authorized MLOps engineers can deploy new models to production. Guard against model inversion attacks (where an attacker queries the model to reconstruct sensitive training data) and model extraction attacks (where an attacker copies the model's functionality by probing it with many queries).

3. OT/IT Convergence Security: The bridge between the factory floor (OT) and the cloud (IT) is a critical vulnerability.

Zero-Trust Architecture: Assume no entity, inside or outside the network, is trusted. Enforce strict identity verification, micro-segmentation of networks (isolating the packaging line network from the R&D network), and least-privilege access for all data flows.

Hardened Edge Devices: Secure the industrial PCs and gateways running edge AI with dedicated security software, regular patches, and physical tamper detection.

7.3.2 Compliance with Evolving Regulations: Navigating a Complex Landscape

The regulatory environment is struggling to keep pace with technology, creating a complex patchwork of rules governing data and automated decisions.

Expanded Compliance Analysis:

1. Data Privacy Regulations (GDPR, CCPA, et al.): While focused on personal data, these laws have implications for industrial AI.

    Data Subject Rights: If sensor data can be linked to an individual worker (e.g., productivity analytics), they may have rights to access, correct, or delete that data. Anonymization and aggregation strategies are crucial.

    Lawful Basis for Processing: Companies must document the lawful basis (e.g., "legitimate interest" for predictive maintenance) for collecting and using data.

    Data Protection by Design: Privacy safeguards must be integrated into the design of AI systems and data pipelines from the outset.

2. Emerging AI-Specific Legislation: The EU AI Act and similar proposals worldwide introduce a risk-based regulatory framework.

   High-Risk AI Systems: Many industrial AI applications (safety components of machinery, critical infrastructure management) will likely be classified as "high-risk." This mandates strict requirements for risk management, data governance, technical documentation, human oversight, and robustness/accuracy.

   Transparency Obligations: Regulations will mandate that users be informed when they are interacting with an AI system and be provided with clear, meaningful explanations of automated decisions that significantly affect them.

3. Industry-Specific Compliance: Manufacturers must continue to meet existing standards (e.g., ISO 9001 for quality, IEC 62443 for OT security) while demonstrating how their AI systems satisfy these requirements. This means creating new, hybrid audit trails that document both the physical process and the digital decision logic that guided it.

Conclusion: Building the Responsible Cognitive Factory

Implementing AI in manufacturing is an exercise in managing powerful dualities: efficiency and ethics, autonomy and accountability, innovation and precaution. This chapter has argued that navigating the risks of bias, opacity, and insecurity is not a separate compliance task but is intrinsic to the engineering of robust, valuable, and enduring cognitive systems. By institutionalizing bias mitigation through diverse teams and continuous auditing, demanding explainability as a core feature of mission-critical AI, and adopting a "security-first" defense-in-depth strategy for data and models, manufacturers build more than just smart factories—they build trustworthy, resilient, and responsible ones. The ultimate competitive advantage in the cognitive age will belong to those who master not only the science of artificial intelligence but also the wisdom required to harness it for the sustainable benefit of their workforce, their customers, and society at large.

CHAPTER 8: THE FUTURE FRONTIER AND CONTINUOUS REINVENTION

Introduction: From Transformation to Perpetual Evolution

The journey through the cognitive factory is not a linear path with a definitive endpoint, but an ascent into a new operational paradigm characterized by perpetual learning and adaptation. This handbook has provided the foundational frameworks for strategy, human augmentation, implementation, and governance. Yet, the landscape upon which these principles are applied is not static. The final chapter peers over the horizon to examine the next wave of technological catalysts and, more importantly, articulates the essential organizational mindset required to harness them. The ultimate differentiator in the coming decade will not be the sophistication of any single algorithm, but the institutional capacity for continuous reinvention. This chapter explores the future frontier of generative design and sustainable innovation, and it culminates in a blueprint for cultivating the agile, learning organization—the only entity capable of thriving in an age of accelerating, AI-driven change (Senge, 2006; Iansiti & Lakhani, 2020).

8.1 Emerging Technologies on the Horizon: The Next Catalysts of Disruption

While current AI applications focus predominantly on optimizing *existing* processes, the next frontier involves AI as a co-creator in the invention of *new* products, materials, and systems. These technologies will further compress innovation cycles and redefine relationships across the value chain.

8.1.1 Generative AI for Industrial Design & Development: The Partner as Co-Creator

Generative AI models, such as Large Language Models (LLMs) like GPT and multimodal systems like DALL-E and Stable Diffusion, are transitioning from tools for content creation to engines of industrial design. This shift fundamentally alters the product lifecycle's earliest stages, creating both immense opportunity and new pressures for manufacturers (Brynjolfsson & McAfee, 2017).

Expanded Impact Analysis:

1. Accelerated Concept Generation & Feasibility Testing: Brand partners and internal R&D teams will use natural language prompts to generate thousands of novel product concepts, packaging designs, or component geometries in hours instead of months. A

prompt such as, "Generate 50 ergonomic handle designs for a cordless drill that reduce wrist strain by 30%, considering common grip types," can yield a diverse portfolio of AI-rendered 3D models. For the manufacturer, this means:

> *Shorter Time-to-Market:* The conceptual phase is compressed dramatically.
>
> *Increased Complexity in Collaboration:* Manufacturers will receive not just finalized specifications, but streams of AI-generated concepts for rapid prototyping and feedback. This requires new digital collaboration platforms and the capability for rapid manufacturability analysis.
>
> *Democratization of Design:* Marketing teams and even end-users may contribute to the generative process, flooding manufacturers with highly customized, *long-tail* product ideas that challenge traditional mass-production economics.

2. Generative Engineering & Topology Optimization 2.0: Beyond aesthetics, generative design algorithms will evolve from creating structurally efficient shapes within set constraints to *understanding and proposing constraints themselves.* An AI could be given a high-level goal: "Design a bracket that supports 50kg, uses

minimal material, and can be manufactured via both metal 3D printing and CNC machining for supply chain resilience." The AI would not only generate the geometry but also propose alternative material choices, compare cost/timeline trade-offs between the two processes, and simulate performance under dynamic stress conditions. This moves generative design from a CAD plugin to a core strategic tool for concurrent engineering.

3. AI-Augmented Simulation and Digital Twin Fidelity: Generative models will create ultra-high-fidelity synthetic data to train and validate other AI systems. To test a visual inspection AI for a new, physically non-existent product, a generative model can create photorealistic images of that product with every conceivable defect under varying lighting and orientation. Furthermore, Large Language Models (LLMs) will act as natural language interfaces to complex simulation software, allowing engineers to ask, "Simulate the thermal stress on this turbine blade during a cold start in Denver's climate," and receive not just data, but a summarized analysis and recommended design adjustments (Hao, 2023).

8.1.2 AI for Sustainable Innovation: Engineering the Circular Economy

The existential imperative of sustainability will be a primary driver for AI innovation, moving beyond operational efficiency to enabling systemic redesign of products and processes for circularity (Bocken & Geradts, 2020).

Expanded Applications and Implications:

1. AI for Material Discovery and Formulation: The search space for new, sustainable materials (e.g., biodegradable polymers, low-carbon cement alternatives, novel battery electrolytes) is vast and multidimensional. AI, particularly through techniques like Bayesian optimization and generative molecular design, can predict the properties of hypothetical material combinations, virtually screen millions of candidates, and guide lab synthesis toward the most promising leads (Raccuglia et al., 2016). For a chemical manufacturer, this could mean discovering a new catalyst that reduces energy consumption in a primary reaction by 40%, fundamentally altering the environmental footprint of an entire product line.

2. Design for Disassembly, Remanufacturing, and Recycling (DRR): AI will be deployed at the design

stage to ensure end-of-life value recovery. A generative design system will have objectives beyond performance and cost:

> *Disassembly Score:* Maximize the ease of non-destructive separation of components.
>
> *Material Purity Score:* Minimize the use of inseparable material composites that contaminate recycling streams.
>
> *Remanufacturing Potential:* Design critical wear components for easy refurbishment and re-certification.
>
> *An AI could analyze a design for a consumer appliance and recommend:* "Replace the glued polymethyl methacrylate faceplate with a snap-fit high-density polyethylene design. This change increases disassembly time by 12 seconds but raises the material recycling purity from 65% to 94% and allows for faceplate remanufacturing."

3. Dynamic Lifecycle Assessment (LCA) and Supply Chain Optimization: AI will power real-time, granular LCAs. By integrating data from IoT sensors across the supply chain—energy mix at supplier factories, logistics fuel consumption, in-plant resource use—an

AI can provide a continuously updated carbon and environmental footprint for each individual product unit. This enables:

Carbon-Aware Production Scheduling: The AI scheduler prioritizes production runs when the local grid's renewable energy percentage is highest.

Sustainable Supplier Selection: Dynamic scoring of suppliers not just on cost and quality, but on their real-time environmental performance data.

Transparent Consumer Footprinting: Providing verifiable, asset-level sustainability credentials to meet regulatory and consumer demand.

8.2 The Agile, Learning Organization: The Ultimate Competitive Advantage

Technology is a tool; culture is the engine. The most advanced AI stack will fail in a rigid, hierarchical, and risk-averse organization. The final, and most critical, frontier is the cultivation of an organizational phenotype built for continuous learning and adaptation—an agile, learning organization (Edmondson, 2018).

8.2.1 Embracing the Mindset of Perpetual Beta

This requires institutionalizing the principles of psychological safety, iterative experimentation, and networked learning that were introduced in earlier chapters.

Expanded Cultural and Structural Framework:

1. From Silos to Fluidity: Reorganizing for Flow: The traditional functional silo (R&D, Engineering, Production, Maintenance) is a barrier to AI-driven innovation. The learning organization adopts fluid, cross-functional "stream teams" organized around value streams or product families. A team might include a design engineer, a data scientist, a production operator, a quality specialist, and a sustainability officer, all empowered to run experiments, from a small process tweak to the pilot of a new generative design tool. Decision-making is pushed to the edges, where the information and context reside (Rigby et al., 2020).

2. Institutionalizing Experimentation: Learning becomes a measurable output, not a happy accident.

    The Corporate "Sandbox": Establish secure, sanctioned digital and physical environments (a pilot line, a cloud instance with synthetic data) where teams can test new AI models and workflows with minimal bureaucratic overhead.

Metrics that Reward Learning: Supplement traditional KPIs (output, yield) with innovation metrics: *Hypotheses Tested per Quarter, Cost of Learning* (funds spent on failed experiments that yielded valuable insights), *Time from Idea to Pilot.* Celebrate "intelligent failures" that provide decisive learning as publicly as successful projects (Edmondson, 2018).

3. Leadership as Context-Setter and Enabler: Leaders in the learning organization shift from being top-down commanders to becoming architects of context. Their primary roles are to:

Articulate a Compelling "Why": Continuously connect the work of experimentation to the organization's broader mission of sustainability, customer value, and societal impact.

Remove Systemic Friction: Actively identify and dismantle policies, IT systems, or budgetary practices that slow down experimentation and learning.

Model the Mindset: Publicly engage in their own learning journeys, acknowledge their knowledge gaps, and demonstrate curiosity (Senge, 2006).

8.2.2 The Final Synthesis: A Blueprint for Action

The age of AI-driven manufacturing is not a future state to be awaited; it is a present reality to be navigated with deliberate action. This handbook culminates in a final, integrated recommendation—a blueprint to begin, sustain, and lead this journey.

The Cognitive Factory Action Framework:

1. **Start with Data, Not Algorithms:** Your most immediate asset is the data you already generate. Begin today by auditing your data landscape. Identify one critical process—a packaging line, a heat treat operation, a quality inspection point—and instrument it to capture clean, structured, and contextualized data. This foundational step enables everything that follows (Davenport, 2018).

2. **Empower Your People as Architects, Not Passengers:** The transformation will be carried out by your workforce or it will fail. Immediately invest in building foundational digital literacy at all levels. Identify and support your "digital champions" on the

floor. Involve operators and technicians in the very first design sessions for AI tools. Their contextual expertise is the irreplaceable component that turns a smart system into a wise one (Daugherty & Wilson, 2018).

3. Think in Pilots, Scale with Platforms: Abandon the quest for a single, monolithic "AI transformation." Adopt a portfolio mindset. Launch small, focused pilot projects with clear success metrics, using the phased framework from Chapter 6. Each pilot, whether a success or an intelligent failure, generates the learning and confidence necessary for the next. Concurrently, build the enabling platforms—the data infrastructure, the MLOps capabilities, the Center of Excellence—that will allow successful pilots to scale efficiently (Fountaine et al., 2019).

4. Govern with Foresight, Not Just Oversight: From day one, integrate the principles of responsible AI. Bake fairness checks, explainability requirements, and security protocols into your development lifecycle. Engage with legal and compliance teams proactively to navigate the evolving regulatory landscape. Trust, once lost, is extraordinarily difficult to regain (Raji et al., 2020).

5. Reinvent Relentlessly: Finally, recognize that this is not a project with an end date. It is the new core competency of your enterprise. Cultivate the learning organization. Stay relentlessly curious about emerging technologies like generative AI and sustainable innovation. View every process, every product, and every business model as temporary and subject to reinvention (Iansiti & Lakhani, 2020).

Conclusion: The Never-Ending Ascent

The cognitive factory is not a destination on a map, but a direction of travel—a commitment to perpetual evolution guided by intelligence, both human and artificial. This handbook has provided the compass, the tools, and the maps for this journey: a strategy rooted in augmentation, a workforce redefined by symbiosis, a pragmatic implementation framework, and an ethical foundation for trust. The frontier ahead, shaped by generative creation and sustainable imperative, is vast and promising.

Your task now is to begin. Start where you are. Use what you have. Do what you can. The complexity is manageable not through a single grand plan, but through a series of deliberate, learning-focused steps. The future of manufacturing will be built not by those who merely adopt technology, but by those who build the organizations capable of learning, adapting, and reinventing alongside it. The ascent continues. Begin your climb today.

REFERENCE

Amiji, M. M., Vlerken, L. E., Yadav, S., & Little, S. R. (2009). Evaluations of combination MDR-1 gene silencing and paclitaxel administration in biodegradable polymeric nanoparticle formulations to overcome multidrug resistance in cancer cells. Cancer Chemotherapy and Pharmacology, 63(4), 711–722.

Anderson, R. (2020). Security engineering: A guide to building dependable distributed systems (3rd ed.). Wiley.

Azadeh, K., De Koster, R., & Roy, D. (2019). Robotized and automated warehouse systems: Review and recent developments. Transportation Science, 53(4), 917–945.

Balbus, J. M., Maynard, A. D., Colvin, V. L., Castranova, V., Daston, G. P., Denison, R. A., Dreher, K. L., Goering, P. L., Goldberg, A. M., Kulinowski, K. M., Monteiro-Riviere, N. A., Oberdörster, G., & Omenn, G. S. (2007). Meeting Report: Hazard assessment for nanoparticles—report from an interdisciplinary workshop. Environmental Health Perspectives, 115(11), 1654–1659.

Benn, T. M., & Westerhoff, P. (2008). Nanoparticle silver released into water from commercially available sock fabrics. Environmental Science & Technology, 42(11), 4133–4139.

Blumer, H. (1966). The idea of social development. Studies in Comparative International Development, 2(1), 3–11.

Blumer, H., Maines, D. R., & Morrione, T. J. (1990). Industrialization as an agent of social change: A critical analysis. Aldine de Gruyter.

Bocken, N. M. P., & Geradts, T. H. J. (2020). Barriers and drivers to sustainable business model innovation: Organization design and dynamic capabilities. Long Range Planning, 53(4), 101950.

Bostrom, N. (2014). Superintelligence: Paths, dangers, strategies. Oxford University Press.

Brode, B. (2022, March 21). AI and nanotechnology are working together to solve real-world problems. Stack Overflow Blog. https://stackoverflow.blog/2022/03/21/ai-and-nanotechnology-are-working-together-to-solve-real-world-problems/

Brynjolfsson, E., & McAfee, A. (2014). The second machine age: Work, progress, and prosperity in a time of brilliant technologies. W. W. Norton & Company.

Brynjolfsson, E., & McAfee, A. (2017). Machine, platform, crowd: Harnessing our digital future. W.W. Norton & Company.

Business Insider. (2024, March 6). ChatGPT may be coming for our jobs. Here are the 10 roles that AI is most likely to replace. https://www.businessinsider.com/chatgpt-jobs-at-risk-replacement-artificial-intelligence-ai-labor-trends-2023-02

Choi, T. M., Wallace, S. W., & Wang, Y. (2021). Big data analytics in operations management. Production and Operations Management, 30(6), 1557–1564.

Dalton, J. (1803). A new system of chemical philosophy. Manchester Literary and Philosophical Society.

Daugherty, P. R., & Wilson, H. J. (2018). Human + machine: Reimagining work in the age of AI. Harvard Business Review Press.

Davenport, T. H. (2018). The AI advantage: How to put the artificial intelligence revolution to work. MIT Press.

DeAngelis, P. L. (2009). Hyaluronan synthases: Fascinating glycosyltransferases from vertebrates, bacterial pathogens, and algal viruses. Cellular and Molecular Life Sciences, 66(11-12), 1923–1939.

Deloitte Insights. (2025). Tech trends 2026. https://www2.deloitte.com/us/en/insights/focus/tech-trends.html

Deng, L., Wang, D., Zhang, Z., & Li, H. (2019). Deep learning-based visual inspection in manufacturing: A survey. Journal of Intelligent Manufacturing, 30(8), 2975–2995. https://doi.org/10.1007/s10845-019-01517-5

De Sousa Jabbour, A. B. L., Jabbour, C. J. C., Godinho Filho, M., & Roubaud, D. (2018). Industry 4.0 and the circular economy: A proposed research agenda and original roadmap for sustainable operations. Annals of Operations Research, 270(1-2), 273–286.

Dias, J., Lima, T., & Paulo Silva, J. (2020). Automated visual inspection in the beverage industry: A review. Machines, 8(2), 20.

Doerr, J. (2018). Measure what matters: How Google, Bono, and the Gates Foundation rock the world with OKRs. Portfolio/Penguin.

Donaldson, K., & Seaton, A. (2005). The hazards of nanoparticles. Occupational Medicine, 55(5), 305–306.

Douglas, S. M., Dietz, H., Liedl, T., Högberg, B., Graf, F., & Shih, W. M. (2009). Self-assembly of DNA into nanoscale three-dimensional shapes. Nature, 459(7245), 414–418.

Drexler, K. E. (1986). Engines of creation: The coming era of nanotechnology. Anchor Books.

Drexler, K. E. (1992). Nanosystems: Molecular machinery, manufacturing, and computation. John Wiley & Sons.

Drexler, K. E., Peterson, C., & Pergamit, G. (1991). Unbounding the future: The nanotechnology revolution. William Morrow.

Edmondson, A. C. (2018). The fearless organization: Creating psychological safety in the workplace for learning, innovation, and growth. John Wiley & Sons.

Egbuna, C., Mishra, A. P., & Goyal, M. R. (Eds.). (2021). Nano- and micro-scale drug delivery systems: Design and fabrication. Elsevier.

Egbuna, C., Parmar, V. K., Jeevanandam, J., Ezzat, S. M., Patrick-Iwuanyanwu, K. C., Adetunji, C. O., ... & Onyeike, E. N. (2021). Toxicity of nanoparticles in biomedical application: Nanotoxicology. Journal of Toxicology, 2021, 9954443.

Eigler, D. M., & Schweizer, E. K. (1990). Positioning single atoms with a scanning tunnelling microscope. Nature, 344(6266), 524–526.

Elder, A., Gelein, R., Silva, V., Feikert, T., Opanashuk, L., Carter, J., ... & Oberdörster, G. (2006). Translocation of inhaled ultrafine manganese oxide particles to the central nervous system. Environmental Health Perspectives, 114(8), 1172–1178.

Fang, H. (2015). Managing data lakes in big data era: What's a data lake and why has it become popular in data management? 2015 IEEE International Conference on Cyber Technology in Automation, Control, and Intelligent Systems (CYBER), 820–824. IEEE.

Feynman, R. P. (1960). There's plenty of room at the bottom. Engineering and Science, 23(5), 22–36.

Feynman, R. P., & Feynman, M. (2006). Classic Feynman: All the adventures of a curious character. W. W. Norton & Company.

Feynman, R. P., & MacCann, G. (2000). The pleasure of finding things out: The best short works of Richard P. Feynman. Basic Books.

Fountaine, T., McCarthy, B., & Saleh, T. (2019). Building the AI-powered organization. Harvard Business Review, 97(4), 62–73.

Fozdar, D. Y., Wu, X., Patrick Jr, C. W., & Chen, S. (2008). Micro-well texture printed into PEG hydrogels using the FILM nanomanufacturing process affects the behavior of preadipocytes. Biomedical Microdevices, 10(6), 839–849.

Fritz, S. (2002). Nanotechnology: Introduction, applications, and implications. U.S. Environmental Protection Agency.

Garbuio, M., & Lin, N. (2019). Artificial intelligence as a growth engine for health care startups: Emerging business models. California Management Review, 61(2), 59–83.

Glenn, J. C., & Gordon, T. J. (2006). Update on the state of the future. The Futurist, 40(1), 20–24.

Gray, T. (2009). The elements: A visual exploration of every known atom in the universe. Black Dog & Leventhal Publishers.

Greenaway, F. (1966). John Dalton and the atom. Cornell University Press.

Gundupalli, S. P., Hait, S., & Thakur, A. (2017). A review on automated sorting of source-separated municipal solid waste for recycling. Waste Management, 60, 56–74.

Gunning, D., Stefik, M., Choi, J., Miller, T., Stumpf, S., & Yang, G. Z. (2019). XAI—Explainable artificial intelligence. Science Robotics, 4(37), eaay7120.

Hao, K. (2023, March 8). AI is dreaming up drugs that no one has ever seen. Now we've got to see if they work. MIT Technology Review.

Hede, S. (2007). Molecular materials and its technology: Disruptive impact on industrial and socio-economic areas. AI & Society, 21(3), 303–313.

Horner, D. S. (2005). Anticipating ethical challenges: Is there a coming era of nanotechnology? Ethics and Information Technology, 7(3), 127–133.

Huang, Y., & Li, X. (2022). The "Amazon effect" and supply chain transformation: A literature review. Journal of Business Research, 142, 1092–1105.

Iansiti, M., & Lakhani, K. R. (2020). Competing in the age of AI: Strategy and leadership when algorithms and networks run the world. Harvard Business Review Press.

Ivanov, D. (2022). Introduction to supply chain resilience: Management, modelling, technology. Springer.

Ivanov, D., Dolgui, A., & Sokolov, B. (2022). Cloud supply chain: Integrating Industry 4.0 and digital platforms in the "Supply Chain-as-a-Service". Transportation Research Part E: Logistics and Transportation Review, 160, 102676.

Jarmon, L., Keating, E., & Toprac, P. (2008). Examining the societal impacts of nanotechnology through simulation: Nano scenario. Simulation & Gaming, 39(2), 168–183.

Jardine, A. K. S., Lin, D., & Banjevic, D. (2006). A review on machinery diagnostics and prognostics implementing condition-based maintenance. Mechanical Systems and Signal Processing, 20(7), 1483–1510. https://doi.org/10.1016/j.ymssp.2005.09.012

John, A. A., Wadhwa, S., & Mathur, A. (2022). Nanotoxicology: An emerging discipline. In Nanomaterials and environmental biotechnology (pp. 239–258). Springer.

Juran, J. M., & Godfrey, A. B. (1999). Juran's quality handbook (5th ed.). McGraw-Hill.

Kagermann, H., Helbig, J., Hellinger, A., & Wahlster, W. (2013). Recommendations for implementing the strategic initiative INDUSTRIE 4.0: Securing the future of German manufacturing industry; final report of the Industrie 4.0 Working Group. Forschungsunion.

Kobbacy, K. A. H., & Murthy, D. N. P. (2008). Complex system maintenance handbook. Springer.

Koenig Solutions. (2023, May 5). Top 19 new technology trends emerging in 2023. https://www.koenig-solutions.com/blog/top-new-technology-trends

Kourti, T. (2005). Application of latent variable methods to process control and multivariate statistical process control in industry. International Journal of Adaptive Control and Signal Processing, 19(4), 213–246. https://doi.org/10.1002/acs.859

Kotter, J. P. (2012). Leading change. Harvard Business Review Press.

Kurzweil, R. (2005). The singularity is near: When humans transcend biology. Viking.

LeCun, Y., Bengio, Y., & Hinton, G. (2015). Deep learning. Nature, 521(7553), 436–444.

Lei, Y., Li, N., Guo, L., Li, N., Yan, T., & Lin, J. (2018). Machinery health prognostics: A systematic review from data acquisition to RUL prediction. Mechanical Systems and Signal Processing, 104, 799–834. https://doi.org/10.1016/j.ymssp.2017.11.016

Li, N., Sioutas, C., Cho, A., Schmitz, D., Misra, C., Sempf, J., ... & Nel, A. (2003). Ultrafine particulate pollutants induce oxidative stress and mitochondrial damage. Environmental Health Perspectives, 111(4), 455–460.

Li, S., Xu, L. D., & Zhao, S. (2021). 5G Internet of Things: A survey. Journal of Industrial Information Integration, 23, 100257.

Liu, B. (2012). Sentiment analysis and opinion mining. Synthesis Lectures on Human Language Technologies, 5(1), 1–167.

Matheson, E., Minto, R., Zampieri, E. G., Faccio, M., & Rosati, G. (2022). Human-robot collaboration in manufacturing applications: A review. Robotics, 11(1), 22.

McCarthy, J., Minsky, M. L., Rochester, N., & Shannon, C. E. (1955). A proposal for the Dartmouth Summer Research Project on Artificial Intelligence. Dartmouth College.

McKinsey & Company. (2025). McKinsey technology trends outlook 2025. https://www.mckinsey.com/capabilities/mckinsey-digital/our-insights/the-top-trends-in-tech

Merolla, P. A., Arthur, J. V., Alvarez-Icaza, R., Cassidy, A. S., Sawada, J., Akopyan, F., ... & Modha, D. S. (2014). A million spiking-neuron integrated circuit with a scalable communication network and interface. Science, 345(6197), 668–673.

Mitchell, M., Wu, S., Zaldivar, A., Barnes, P., Vasserman, L., Hutchinson, B., ... & Gebru, T. (2019). Model cards for model reporting. In Proceedings of the conference on fairness, accountability, and transparency (pp. 220–229).

Moore, W. E. (1967). Order and change: Essays in comparative sociology. John Wiley & Sons.

Mourtzis, D. (2020). Simulation in the design and operation of manufacturing systems: State of the art and new trends. International Journal of Production Research, 58(7), 1927–1949. https://doi.org/10.1080/00207543.2019.1636321

Mourtzis, D., Angelopoulos, J., & Panopoulos, N. (2021). A literature review of the challenges and opportunities of the transition from Industry 4.0 to Society 5.0. Energies, 14(6), 1596.

Nanotoxicology. (2023). In Wikipedia. Retrieved [Date you accessed it] from https://en.wikipedia.org/wiki/Nanotoxicology

National Academies of Sciences, Engineering, and Medicine. (2018). Reducing the threat of chemical, biological, radiological, and nuclear terrorism. The National Academies Press. https://doi.org/10.17226/25174

National Nanotechnology Initiative. (2021). Nanotechnology: Big things from a tiny world. U.S. National Science and Technology Council.

National Nanotechnology Initiative. (2021, October). 2021 National Nanotechnology Initiative strategic plan. https://www.nano.gov/sites/default/files/pub_resource/NNI-2021-Strategic-Plan.pdf

National Nanotechnology Initiative (NNI). (n.d.). About the NNI. Retrieved [Date you accessed it] from https://www.nano.gov/about-nni

O'Neil, C. (2016). Weapons of math destruction: How big data increases inequality and threatens democracy. Crown.

Oberdörster, G., Oberdörster, E., & Oberdörster, J. (2005). Nanotoxicology: An emerging discipline evolving from studies of ultrafine particles. Environmental Health Perspectives, 113(7), 823–839.

Otto, B. (2011). Data governance. Business & Information Systems Engineering, 3(4), 241–244.

Palmberg, C. (2008). The transfer and commercialisation of nanotechnology: A comparative analysis of university and company researchers. Journal of Technology Transfer, 33(6), 631–652.

Paramecium. (n.d.). In Encyclopedia Britannica. Retrieved [Date you accessed it] from https://www.britannica.com/science/Paramecium

Pinedo, M. L. (2016). Scheduling: Theory, algorithms, and systems (5th ed.). Springer.

Porter, M. E., & Heppelmann, J. E. (2017). Why every organization needs an augmented reality strategy. Harvard Business Review, 95(6), 46–57.

Powell, M. C. (2007). New risk or old risk, high risk or no risk? How scientists' standpoints shape their nanotechnology risk frames. Health, Risk & Society, 9(2), 173–190.

Qin, S. J., & Badgwell, T. A. (2003). A survey of industrial model predictive control technology. Control Engineering Practice, 11(7), 733–764. https://doi.org/10.1016/S0967-0661(02)00186-7

Raccuglia, P., Elbert, K. C., Adler, P. D. F., Falk, C., Wenny, M. B., Mollo, A., ... & Norquist, A. J. (2016). Machine-learning-assisted materials discovery using failed experiments. Nature, 533(7601), 73–76.

Raji, I. D., Smart, A., White, R. N., Mitchell, M., Gebru, T., Hutchinson, B., ... & Barnes, P. (2020). Closing the AI accountability gap: Defining an end-to-end framework for internal algorithmic auditing. In Proceedings of the 2020 Conference on Fairness, Accountability, and Transparency (pp. 33–44).

Ribeiro, M. T., Singh, S., & Guestrin, C. (2016). "Why should I trust you?" Explaining the predictions of any classifier. In Proceedings of the 22nd ACM SIGKDD international conference on knowledge discovery and data mining (pp. 1135–1144).

Ries, E. (2011). The lean startup: How today's entrepreneurs use continuous innovation to create radically successful businesses. Crown Business.

Rigby, D. K., Sutherland, J., & Takeuchi, H. (2020). The secret history of agile innovation. Harvard Business Review, 98(2), 84–93.

RNCOS. (2008, April). Nanotechnology market forecast to 2015. RNCOS E-Services Pvt. Ltd.

Romero, D., Stahre, J., Wuest, T., Noran, O., Bernus, P., Fast-Berglund, Å., & Gorecky, D. (2020). Towards an operator 4.0 typology: A human-centric perspective on the fourth industrial revolution technologies. Computers & Industrial Engineering, 139, 105644.

Rothemund, P. W. (2006). Folding DNA to create nanoscale shapes and patterns. Nature, 440(7082), 297–302.

Saxena, S. K., Nyodu, R., Kumar, S., & Maurya, V. K. (2020). Current advances in nanotechnology and medicine. In NanoBioMedicine (pp. 3–16). Springer.

Sculley, D., Holt, G., Golovin, D., Davydov, E., Phillips, T., Ebner, D., ... & Dennison, D. (2015). Hidden technical debt in machine learning systems. In Advances in neural information processing systems 28 (pp. 2503–2511).

Seeman, N. C. (2010). Nanomaterials based on DNA. Annual Review of Biochemistry, 79, 65–87.

Selcuk, S. (2017). Predictive maintenance, its implementation and latest trends. Proceedings of the Institution of Mechanical Engineers, Part B: Journal of Engineering Manufacture, 231(9), 1670–1679. https://doi.org/10.1177/0954405415601640

Senge, P. M. (2006). The fifth discipline: The art & practice of the learning organization (Revised ed.). Currency Doubleday.

Soares, S., Sousa, J., Pais, A., & Vitorino, C. (2018). Nanomedicine: Principles, properties, and regulatory issues. Frontiers in Chemistry, 6, 360. https://doi.org/10.3389/fchem.2018.00360

Stang, C., & Sheremeta, L. (2006). Nanotechnology—a lot of hype over almost nothing? Health Law Review, 15(1), 53–55.

Stouffer, K., Pillitteri, V., Lightman, S., Abrams, M., & Hahn, A. (2015). Guide to industrial control systems (ICS) security (NIST Special Publication 800-82, Rev. 2). National Institute of Standards and Technology.

TechJury. (2024, September 11). New technology trends for 2024: Exploring digital breakthroughs. TechJury. https://techjury.net/blog/new-technology-trends/

The White House, Office of Science and Technology Policy. (2021, October). Biden-Harris administration announces new national nanotechnology initiative strategic plan [Press release]. https://www.whitehouse.gov/ostp/news-updates/2021/10/21/biden-harris-administration-announces-new-national-nanotechnology-initiative-strategic-plan/

Toffler, A. (1984). The third wave. Bantam Books.

Turing, A. M. (1950). Computing machinery and intelligence. Mind, 59(236), 433–460.

Uddin, F. (2008). Clays, nanoclays, and montmorillonite minerals. Metallurgical and Materials Transactions A, 39(12), 2804–2814.

Vallance, C. (2023, March 28). AI could replace equivalent of 300 million jobs - report. BBC News. https://www.bbc.com/news/technology-65102150

van der Aalst, W. M. P. (2016). Process mining: Data science in action (2nd ed.). Springer.

Vaswani, A., Shazeer, N., Parmar, N., Uszkoreit, J., Jones, L., Gomez, A. N., ... & Polosukhin, I. (2017). Attention is all you need. Advances in Neural Information Processing Systems, 30, 5998–6008.

Wang, L. (2006). Nanotechnology: A new industrial revolution. Intertech.

Whitesides, G. M., & Grzybowski, B. (2002). Self-assembly at all scales. Science, 295(5564), 2418–2421.

World Economic Forum. (2023). Future of jobs report 2023. https://www.weforum.org/reports/the-future-of-jobs-report-2023/

Wuest, T., Weimer, D., Irgens, C., & Thoben, K. D. (2016). Machine learning in manufacturing: Advantages, challenges, and applications. Production & Manufacturing Research, 4(1), 23–45.

Zhou, K., Liu, T., & Zhou, L. (2019). Industry 4.0: Towards future industrial opportunities and challenges. 2019 12th International Conference on Fuzzy Systems and Knowledge Discovery (FSKD), 2147–2152.

APPENDIX:

GLOSSARY OF KEY AI AND INDUSTRY 4.0 TERMS

A

Advanced Analytics: The application of sophisticated techniques and tools—including machine learning, predictive modeling, and statistical algorithms—to data to discover deeper insights, make predictions, or generate recommendations, going beyond traditional business intelligence (descriptive analytics).

Algorithm: A finite sequence of well-defined, computer-implementable instructions, typically to solve a class of problems or to perform a computation. In AI, algorithms are the procedural rules that govern how a model learns from data.

Artificial Intelligence (AI): The broad field of computer science focused on creating systems capable of performing tasks that typically require human intelligence. These tasks include learning, reasoning, problem-solving, perception, and natural language understanding. **Manufacturing AI** refers specifically to the application of these technologies within industrial operations.

Augmented Reality (AR): A technology that superimposes computer-generated information (images, text, animations) onto a user's view of the real world, typically through smart glasses or tablet/smartphone screens. In manufacturing, AR is used for guided assembly, maintenance, training, and remote assistance.

Augmented Workforce: A strategic approach where AI and automation tools are designed to amplify human capabilities rather than replace them. The human worker acts as an orchestrator and decision-maker, leveraging AI for data analysis, repetitive tasks, and pattern recognition.

Automation Bias: The human tendency to favor suggestions from automated decision-making systems and to ignore contradictory information made without automation, even when the automated system is wrong. This is a critical human factor risk in human-AI collaboration.

B

Big Data: Extremely large and complex datasets that cannot be easily managed, processed, or analyzed with traditional data processing tools. In Industry 4.0, big data is characterized by the **Volume, Velocity, Variety, Veracity, and Value** of data generated by connected machines and sensors.

Black Box Problem: A characteristic of some complex AI models (particularly deep learning) where the internal logic by which inputs are transformed into outputs is opaque and not easily interpretable by humans. This poses challenges for trust, debugging, and regulatory compliance.

C

Cognitive Manufacturing: An advanced stage of smart manufacturing where systems leverage AI, machine learning, and contextual awareness to simulate human thought processes in complex decision-making. It represents a factory that can perceive, learn, reason, and assist in problem-solving.

Computer Vision: A field of AI that enables computers to derive meaningful information from digital images, videos, and other visual inputs, and to take actions or make recommendations based on that information. In manufacturing, it is used for automated visual inspection, defect detection, and robotic guidance.

Cyber-Physical System (CPS): An engineered system that seamlessly integrates computational algorithms and physical components. Sensors and actuators bridge the cyber (digital) and physical worlds, enabling real-time data collection, analysis, and control of physical processes. The smart factory is a network of CPS.

D

Digital Thread: A communication framework that connects data flows and provides an integrated view of an asset's data (e.g., a product, machine, or process) throughout its lifecycle—from design and manufacturing to service and decommissioning. It ensures traceability and context.

Digital Twin: A virtual, dynamic representation of a physical object, system, or process. It uses real-time data from sensors and other sources to simulate, predict, and optimize performance. **Process Twins** model workflows, while **Asset Twins** model physical equipment.

E

Edge Computing: A distributed computing paradigm that brings computation and data storage closer to the location where it is needed (i.e., the "edge" of the network, such as on the factory floor). This reduces latency, conserves bandwidth, and allows for real-time processing critical for AI applications like robotic control.

Enterprise Resource Planning (ERP): A software system used to manage and integrate the finance, manufacturing, and operations in an organization.

Explainable AI (XAI): A set of methods and techniques in AI that allows human users to understand, trust, and effectively manage the results created by machine learning models. It addresses the "black box" problem by making model decisions interpretable.

I

IEC 62443: A series of international standards for securing industrial automation and control systems (IACS), addressing their unique needs for uptime and safety.

Industrial Internet of Things (IIoT): The network of physical industrial devices (sensors, actuators, machines, controllers) embedded with connectivity, enabling them to collect, monitor, exchange, and analyze data. IIoT is the foundational infrastructure for data collection in Industry 4.0.

Industry 4.0: The ongoing automation of traditional manufacturing and industrial practices using modern smart technology, characterized by the fusion of the physical, digital, and biological worlds. It encompasses CPS, IIoT, cloud computing, cognitive computing, and AI to create the "smart factory."

L

Lights-Out Manufacturing: A production model with full automation where human intervention is not required, allowing facilities to operate in the dark. While often aspirational, the modern goal is a highly automated factory where human workers supervise and manage exceptions, not routine tasks.

M

Machine Learning (ML): A subset of AI that provides systems the ability to automatically learn and improve from experience without being explicitly programmed. It focuses on the development of algorithms that can access data and use it to learn for themselves. Key types include:

> **Supervised Learning:** The model is trained on labeled data (inputs with known outputs).
>
> **Unsupervised Learning:** The model finds patterns in unlabeled data.

Reinforcement Learning: The model learns by interacting with an environment, receiving rewards or penalties for actions.

Manufacturing Execution System (MES): A computerized system used in manufacturing to track and document the transformation of raw materials to finished goods. It provides real-time data on production orders, inventory, equipment status, and labor to optimize shop floor operations.

Message Queuing Telemetry Transport (MQTT): A lightweight messaging protocol for M2M (machine-to-machine) communication

Model Drift: The degradation of a machine learning model's predictive performance over time due to changes in the underlying real-world data or relationships. **Concept drift** refers to changes in the statistical properties of the target variable, while **data drift** refers to changes in the input data distribution.

Multimodal system: A computing or artificial intelligence (AI) system designed to process, interpret, and integrate information from multiple communication modalities (types of data).

N

Natural Language Processing (NLP): A branch of AI that gives machines the ability to read, understand, and derive meaning from human language. In the factory, NLP powers voice-activated controls, chatbots for operational support, and automated report generation from spoken or written logs.

O

Open Platform Communications (OPC): A set of non-proprietary standards for secure, reliable data exchange in industrial automation, enabling different vendors' devices (PLCs, sensors) and software (SCADA, HMIs) to communicate seamlessly, especially evolving from older OLE-based systems to the modern, platform-independent **OPC UA** (Unified Architecture) that uses TCP/IP for broader connectivity and enhanced security. It acts as a standardized "translator," allowing diverse industrial systems to share real-time data, alarms, and historical data without needing custom drivers, ensuring interoperability

OPC Open Platform Communications -Unified Architecture (OPC UA): The modern, platform-independent (Windows, Linux, macOS, embedded), secure evolution that integrates classic OPC features and adds powerful new capabilities for industrial automation and the IIoT.

Operational Technology (OT): The hardware and software that detects or causes a change through the direct monitoring and/or control of physical devices, processes, and events in the enterprise (e.g., PLCs, SCADA systems, CNC machines). The convergence of OT with IT is a core tenet of Industry 4.0.

Overall Equipment Effectiveness (OEE): A cornerstone metric in manufacturing that measures the productivity of a manufacturing line or asset. It is calculated as: **Availability × Performance × Quality**. AI is used to predict and improve OEE by identifying root causes of loss.

P

Predictive Maintenance (PdM): A maintenance strategy that uses data analysis, sensor information, and machine learning models to predict when an equipment failure is likely to occur, allowing maintenance to be performed just in time. This contrasts with reactive (fix-it-when-it-breaks) or preventive (scheduled) maintenance.

Prescriptive Analytics: The most advanced form of analytics, which goes beyond predicting what will happen and suggests actionable recommendations on what should be done to achieve desired outcomes or avoid future problems. It often employs optimization and simulation algorithms.

Process Mining: A technique designed to discover, monitor, and improve real processes by extracting knowledge from event logs readily available in information systems (ERP, MES). It creates a digital twin of processes to identify bottlenecks, variations, and compliance issues.

Q

Quality Management System (QMS): A formal system that documents processes, procedures, and responsibilities for achieving quality policies and objectives in an organization.

R

Robotic Process Automation (RPA): The use of software "bots" to automate high-volume, repetitive, rule-based digital tasks, such as data entry, form filling, and report generation. In an Industry 4.0 context, RPA often bridges gaps between legacy systems that lack modern APIs.

S

Scope 3 emissions: These are all indirect greenhouse gas (GHG) emissions that occur in a company's value chain, both upstream (suppliers) and downstream (customers), beyond what's covered in Scope 1 (direct) and Scope 2 (purchased energy).

Smart Factory: A highly digitized and connected production facility that leverages technologies from Industry 4.0—such as CPS, IIoT, and AI—to continuously collect and analyze data, enabling flexible, self-optimizing, and efficient production.

Supervisory Control and Data Acquisition (SCADA): A control system architecture comprising computers, networked data communications, and graphical user interfaces for high-level process supervision. It is used to monitor and control plant equipment and infrastructure.

T

Time Series Data: A sequence of data points indexed (or listed) in time order, typically taken at successive, equally spaced points in time. In manufacturing, sensor readings (temperature, vibration, pressure) are classic time series data used for monitoring and predictive analytics.

www.ingramcontent.com/pod-product-compliance
Lightning Source LLC
LaVergne TN
LVHW010201070526
838199LV00062B/4441